BAHAMAS PRIMARY

Mathematics Book 6

The authors and publishers would like to thank the following members of the Teachers' Panel, who have assisted in the planning, content and development of the books:

Chairperson: Dr Joan Rolle, Senior Education Officer, Primary School Mathematics, Department of Education

Team members:

Lelani Burrows, Anglican Education Authority

Deidre Cooper, Catholic Board of Education

LeAnna T. Deveaux-Miller, T.G. Glover Professional Development and Research School

Dr Marcella Elliott Ferguson, University of The Bahamas

Theresa McPhee, Education Officer, High School Mathematics, Department of Education

Joelynn Stubbs, C.W. Sawyer Primary School

Dyontaleé Turnquest Rolle, Eva Hilton Primary School

Karen Morrison, Daphne Paizee and Rentia Pretorius

HODDER EDUCATION
AN HACHETTE UK COMPANY

The Publishers would like to thank the following for permission to reproduce copyright material.

Photo credits

All photos © Mike van der Wolk, Tel: +27 83 2686000, mike@springhigh.co.za, except: page 5 © Rashevskyi Viacheslav/Shutterstock.com; page 18 © RIA Novosti/TopFoto; page 24 © railway fx/ Shutterstock.com; page 34 © Lee Prince/Shutterstock.com; page 93 © creativestockexchange/ Shutterstock.com; page 108 © La Gorda/Shutterstock.com; page 134 © r. roth/123RF; page 153 © National Institute of Standards and Technology.

Orders: please contact Bookpoint Ltd, 130 Park Drive, Milton Park, Abingdon, Oxon OX14 4SE. Telephone: (44) 01235 827720. Fax: (44) 01235 400454. Email education@bookpoint.co.uk Lines are open from 9 a.m. to 5 p.m., Monday to Saturday, with a 24-hour message answering service. You can also order through our website: www.hoddereducation.com

ISBN: 978 1 4718 6474 2

© Cloud Publishing Services 2017

First published in 2017 by
Hodder Education,
An Hachette UK Company
Carmelite House
50 Victoria Embankment
London EC4Y 0DZ

www.hoddereducation.com

Impression number   10 9 8 7 6 5 4 3 2

Year   2021 2020 2019

Cover photo © Thinkstock/iStockphoto/Getty Images

Illustrations by Peter Lubach and Aptara Inc.

Typeset in India by Aptara Inc.

Printed in Slovenia

A catalogue record for this title is available from the British Library.

Contents

Key Words

estimate

calculate

solve

multi-step

▲ You already know that you can find mathematics everywhere you look. Look at the photo carefully. Try to find something in the photo that shows symmetry, a right-angled triangle, parallel lines, intersecting lines, patterns, mass and speed. Share your ideas with your group.

You use mathematics in your daily lives without really thinking about it and you use the skills you learn in school to solve problems that involve **estimating**, **calculating** and finding patterns. In this topic, you are going to revise some of the skills you learned in earlier grades and apply them to **solve** problems like the ones you often come across in real life.

Getting Started

1 The parasailing that you can see in the photograph is a popular beach activity for tourists. It costs $45.00 for a 15 minute trip. The boat operators work from 9:00 a.m. to 4:30 p.m. each day.

   a What is the cost of a trip for two people?

   b What is the maximum amount of money the operators can collect in one day? Show how you worked this out.

2 Make up two word problems using the photograph. One of the problems should be a multi-step problem. You may make up numerical amounts in your problem; for example, you could make up the average mass of a tourist, or the cost of a parasail and so on. Swap problems with another student and try to solve each other's problems. Check the answers in pairs.

3 What are your goals for mathematics this year? Tell your group:

   a what things you hope to improve

   b how you plan to do this.

4 Write your top three goals for this year in your mathematics journal. Next to each one, write one step you will take to reach that goal.

Let's Think …

Antonio ran 200 metres in 24 seconds. Delwyn ran 400 metres in 65 seconds. Leeanne ran 100 metres in 14 seconds. Who is the fastest runner? How did you decide?

You already have enough mathematical skills to complete the activities in this unit. If you have forgotten something or you need to check something, use the contents page to find the correct chapter and read through the information and examples to refresh your memory. You can also ask other students and your teacher for help if you encounter difficulty.

1 Read this short newspaper article.

$3 200.00 Raised for Charity

James Brown, 29, used to weigh 112 kilograms. After a health scare he started to exercise and eat only healthy foods. To encourage him, his friends sponsored his efforts and together they raised $3 200.00 for a children's charity. Today, after 3 months of effort, James is a healthy 82½ kilograms and his waist measurement is an astonishing 0.28 m smaller.

a What do each of the numbers in the article refer to?

b How much weight did James lose?

c If his waist measurement is now 74.5 centimetres, what was it before?

d What is the average number of kilograms that James lost per month?

2 Choose the correct number or operation symbol from the box to complete each calculation.

$\times$	63	0.63	$\div$	630

a $6.3 \times 10 = \square$

b $6.3 \div 10 = \square$

c $3.7 \boxed{} 10 = 37$

d $12.9 \boxed{} 10 = 1.29$

e $0.63 \times 1\,000 = \square$

f $630 \boxed{} 1\,000 = 0.63$

3 Read the information and answer the questions.

a There were n people on a bus. Now there are $n + 4$. What could have happened?

b Micah had m sweets. Now he has $m \div 2$. What could have happened?

c Jerome had x dollars. After his birthday he had $2 \times x + 3$ dollars. What does this mean?

d There are y mangoes in a box. How many would be left if 9 of them were eaten?

4 Use a ruler.

   a Measure the width of this book in millimetres.

   b Measure the length of this book in millimetres.

   c What is the perimeter of a page in this book? Give your answer in centimetres.

5 What units would you use to measure:

   a the thickness of this book?

   b the distance from your classroom to the principal's office?

   c the area of the top of your desk?

   d the area of the Caribbean Sea?

6 Nicole had the bar of chocolate shown in the picture.

She gave 0.25 to Anne and $\frac{3}{8}$ to Cecile. What fraction does she have left?

7 Write each set of measurements in descending order.

a	b	c
2.5	6.35	3.4
1.63	6.53	$3\frac{1}{2}$
0.9	6.035	3.29
1.04	6.05	$3\frac{7}{10}$
3.1	6.5	3.02
3.01	6.03	$3\frac{7}{9}$

8 Joshua is playing a game in which he tosses counters onto a number card to score.

   a List his scores in ascending order.

   b What is the mode of the scores?

   c What is the range of the scores?

   d What is the median score?

●	●	3	4	●
●	●	8	9	10
11	12	●	●	15
16	17	18	●	●

9 Antonique buys a ruler and two pencils at a market stall. The ruler cost $1.50 and the pencils cost 35¢ each. The salesperson works out the total like this:

$150 + 35 \times 2 = 3.70$.

   a Explain why this calculation is wrong.

   b What did the salesperson actually charge Antonique for?

Looking Back

- Which of these activities did you find easiest? Why?
- Was there anything you found difficult? What?

Topic Review

What Did You Learn?

- In this topic you revised some of the things you learned last year.

Talking Mathematics

Answer these questions as mathematically as you can.

- Are all four-sided shapes rectangles?
- What is a survey used for?
- What does it mean if the numerator of a fraction is greater than the denominator?
- Why do square numbers have an odd number of factors?
- Does the word average always mean the same thing?

Quick Check

1   Apply the correct order of operations to work out the answer to the calculations.

   a $6 \times 2 + 3$ b $15 - 10 \div 5$

   c $12 + 8 \times 3$ d $(9 + 1) \times 10$

   e $12 + (6 - 2) \times 3$ f $10 \times (5 + 2 \times 8)$

2   Solve these equations.

   a $x - 2 = 8$ b $p - 8 = 15$

   c $x + 12 = 1$ d $9 \times m = 54$

   e $x \div 10 = 45$ f $m \div 10 = 0.95$

3   A strip of metal 24 cm long is chopped into pieces 5 mm long. How many pieces will there be?

4   It took Naresh 3 hours and 20 minutes to drive to his granny's house. He stopped for 45 minutes for lunch on the way. How long was he driving for?

5   A ferry from Nassau to Paradise Island makes 24 return trips every weekday. On the weekend, it makes 25 return trips on Saturday and 18 on Sunday. How many one-way trips does it make altogether in a week?

6   A group spent $100.00 exactly on entry tickets to an exhibition. It cost $11.00 for each adult and $8.00 for each child. How many children were in the group?

7   Mr Frank lives on Beach Road. There are 50 houses in the road (numbered from 1 to 50). His house number has 2 digits. The sum of the digits is 9 and the product is 14. What is Mr Frank's house number?

Topic 2 Numbers and Place Value Workbook pages 3–5

▲ The number of people on Earth is increasing every second of every day. The United Nations estimated that there were over 7.5 billion people on Earth in 2016 and they think this will increase to 8 billion by 2024. In 2016 around half the people lived in cities and 4.3 billion people lived in Asia. Estimate the city population for 2016 and work out how many people did not live in Asia.

Do you remember all the **place values** to hundred **millions**? Last year, you worked with large **numbers**. You counted, ordered and compared numbers with up to 9 **digits** and you learned about the value of each **place** in the number. This year, you are going to continue to work with large numbers and also extend the place value table to include **billions**.

Getting Started

1 Write these numbers in numerals.
   a Three hundred seven million four hundred ninety-three thousand two hundred twenty-nine.
   b Ninety-five million.

2 Write the number that is:
   a 1 less than 10 000
   b 1 more than 99 999
   c 1 less than 100 000 000.

3 How do you think you write a billion in numerals? Why?

4 Write the number that is halfway between 30 million and 31 million.

5 Say the number that is 1 million less than:
   a 1 234 564   b 432 543 000   c 900 400 600

Unit 1 Revisit Place Value to Hundred Millions

Let's Think …

When would you need to use numbers that are greater than 1 million?
Try to give at least four examples.

Do you remember the places to hundred millions? Read through the information and examples to refresh your memory.

This place value table shows the places from ones to hundred millions.

Hundred Millions	Ten Millions	Millions	Hundred Thousands	Ten Thousands	Thousands	Hundreds	Tens	Ones
3	2	4	4	1	3	0	8	2

To read the number:

- *read the millions first* — *three hundred twenty-four million*
- *read the thousands next* — *four hundred thirteen thousand*
- *read the rest of the number* — *eighty-two*

This number is three hundred twenty-four million four hundred thirteen thousand eighty-two.

You can write it in expanded notation as well like this:

300 000 000 + 20 000 000 + 4 000 000 + 400 000 + 10 000 + 3 000 + 80 + 2

---

1   Say each number aloud. Then write the value of the red digit in each number.

   a 334 000 000   b 210 654 763   c 987 000 987   d 17 234 900   e 999 435 004

2   Write the smaller number in each pair in expanded notation.

   a 645 234 756   647 234 756

   b 812 135 781   702 144 987

   c 734 680 129   437 860 126

3   Write each set of numbers in descending order.

   a 12 983 467   10 203 004   8 747 543   24 302 065   18 765 100

   b 199 046 871   132 098 999   191 098 000   218 021 098   200 987 456

   c 231 432 654   231 876 132   213 312 456   312 342 125   231 987 098

Looking Back

Write in numerals, the number that is:

a   3 million more than 12 500 000   b   10 million more than 2 411 000

c   1 200 000 more than 3 million   d   1 500 000 less than 12 600 000

Arrange your answers in ascending order.

Unit 2 From Millions to Billions

Let's Think ...

- Write the number you think is the greatest nine-digit number possible.
- Say the place and the value of each digit in the number.

The place after hundred millions is called billions.

Billions	Hundred Millions	Ten Millions	Millions	Hundred Thousands	Ten Thousands	Thousands	Hundreds	Tens	Ones
1	0	0	0	0	0	0	0	0	0

You write 1 000 000 000 and say one billion.

One billion is equivalent to one thousand million.

As with the hundreds, thousands and millions, there are three places in the billions group: billions, ten billions and hundred billions.

Billions			Millions			Thousands			Ones		
Hundred Billions	Ten Billions	Billions	Hundred Millions	Ten Millions	Millions	Hundred Thousands	Ten Thousands	Thousands	Hundreds	Tens	Ones
3	2	1	1	5	4	2	1	7	4	3	2

To read the number:

- read the billions first   *three hundred twenty-one billion*
- read the millions next   *one hundred fifty-four million*
- read the thousands next   two hundred seventeen thousand
- read the rest of the number   *four hundred thirty-two*

This number is three hundred twenty-one billion, one hundred fifty-four million, two hundred seventeen thousand, four hundred thirty-two.

---

1 How would you say each of these numbers?

   a 12 345 657   b 1 987 098 432   c 12 000 000

   d 12 000 000 000   e 123 450 000 000   f 312 345 609 300

2 Write down six numbers with 10, 11 or 12 digits. Exchange with a partner and say each other's numbers aloud.

3 Make yourself a foldable place value reminder. You will need a strip of card or thick paper (at least 24 cm long and 5 cm wide), a ruler and some coloured pens.

Follow the instructions:

- Draw a horizontal line across your card.
- Fold the card into quarters and draw a vertical line on the three folds.
- Label the four quarters with the groups: billions, millions, thousands and ones.
- Draw lines to divide the bottom part of each quarter into thirds.
- Label the places in each group correctly.

Billions			Millions			Thousands			Ones		
Hundred Billions	Ten Billions	Billions	Hundred Millions	Ten Millions	Millions	Hundred Thousands	Ten Thousands	Thousands	Hundreds	Tens	Ones

Keep your reminder in your mathematics book and use it as you need to.

4 This table shows the average distance of the planets in the solar system from the Sun. The diagram shows the planets in relation to the Sun. Read and say each number before you start.

Planet	Mean Distance from the Sun (km)
Mercury	57 910 000
Venus	108 200 000
Earth	149 600 000
Mars	227 940 000
Jupiter	778 330 000
Saturn	1 424 600 000
Uranus	2 873 550 000
Neptune	4 501 000 000
Pluto	5 945 900 000

a Use the information in the table to work out which planet is which on the diagram. Show your partner your answer.

b Do you think these are exact distances or estimates? Why?

c Which planets are less than a billion kilometres from the Sun?

d Sharyn says Saturn is almost double the distance from the Sun as Jupiter. Is she correct?

e Shawn says he is thinking of a planet that is about 1 billion kilometres from the Sun. Which two planets could he be thinking of? Why?

f Estimate how much further Pluto is from the Sun than Neptune.

5 The table shows how the world's population changed from 1950 to 2010 and how it is expected to change up to 2050 (the numbers after 2010 are estimates). Read and say each number before you start.

Year	Population
1950	2 557 628 654
1960	3 043 001 508
1970	3 712 697 742
1980	4 444 496 764
1990	5 283 252 948
2000	6 008 279 216
2010	6 856 615 879
2020	7 629 798 111
2030	8 319 146 289
2040	8 897 252 335
2050	9 374 484 225

a In which years was the population less than 5 billion?

b Which is the first year with a population over 6 billion?

c When is the population closest to 9 billion, 2040 or 2050? Give a reason for your answer.

d Look at the population figure for 1980. Write the place and the value of each 4 in the number.

e How long did it take for the 1970 population to double?

f Approximately what would you expect the population to be in 2060? Why?

g What do you think the population was in 1940? Why?

6 Write the following numbers. Compare with a partner once you have finished.

a A number that is greater than 32 billion but less than 32.5 billion.

b A number less than 8 billion with a 7 in the billions place, the hundred thousands place and the ones place. No other sevens may appear in the number.

c The biggest eleven-digit number you can make with five more ten billions than billions and less than 2 billions.

d The smallest number you can make with 8 in the billions place.

7 Use the population figures from 2010 to 2050.

a Write each number backwards in your book.

b Arrange the new numbers in ascending order.

Looking Back

Pluto is a dwarf planet that is 5 945 900 000 km from the Sun.
Write this number in words and in expanded notation.

Topic Review

What Did You Learn?

- You can write any number using the digits from 0 to 9 and place value.
- The place value table extends to the left as numbers get larger.
- Each place is ten times greater than the place to its right.
- You worked with place value to hundred billions.

Billions			Millions			Thousands			Hundreds		
Hundred Billions	Ten Billions	Billions	Hundred Millions	Ten Millions	Millions	Hundred Thousands	Ten Thousands	Thousands	Hundreds	Tens	Ones

- You read the numbers from left to right in groups.
- When you write the numbers, you leave a space between each group of three digits.

Talking Mathematics

How does a good understanding of place value help you read large numbers correctly?

Quick Check

1 Write the greater number in each pair in expanded notation.

   a 12 645 234 756   12 647 234 756   b 312 812 135 781   312 702 144 987

   c 7 734 680 129   7 437 860 126

2 Write each set of numbers in ascending order.

   a 1 312 983 467   1 430 203 004   1 238 747 543   1 324 302 065   1 118 765 100

   b 320 199 046 871   320 132 098 999   320 191 098 000   302 218 021 098   302 200 987 456

   c 99 231 432 654   99 231 876 132   99 213 312 456   99 312 342 125   99 231 987 098

3 China and India are the only two countries in the world with more than a billion people each. Work in groups. Use the table of population data below and prepare a short talk with numbers to present to the rest of the class.

	2016	2050 Expected
China	1 378 561 591	1 303 723 332
India	1 266 883 598	1 656 553 632
Top 10 most populated countries combined	4 249 123 407	4 950 140 178
Rest of the world	3 090 970 573	4 306 202 522

Topic 3 Exploring Patterns Workbook pages 6–7

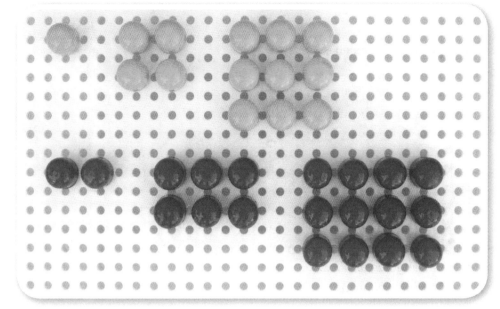

▲ Look at the dot patterns on this pegboard. Do you recognize either of them?
How many pegs would you need to build the next shape in each pattern?
How did you work this out?

You worked with numeric **patterns** as well as **geometric** patterns last year and you know how to describe patterns and work out what comes next. In this topic, you are going to revisit some of these patterns and learn about some new ones. You are also going to investigate and use number patterns to make calculations easier. Mathematics relies on patterns and a good understanding of pattern work can help you solve problems and see relationships between different types of mathematics.

Getting Started

1 The pattern of odd and even numbers is one of the most basic patterns in mathematics.
  a List the odd numbers between 40 and 60.
  b List the even numbers between 100 and 116.
  c Explain how you can tell whether a number in the billions is odd or even.

2 Jay made up this pattern. What should the missing square look like? Why?

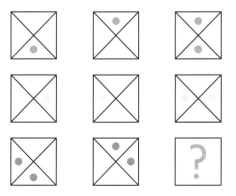

Unit 1 Special Number Patterns

Let's Think …

A square number is the product of multiplying a number by itself.

$3 \times 3 = 9$ So, 3 squared equals 9 and we say that 9 is a square number.

When you draw dot diagrams of square numbers you get a sequence of squares.

These dot diagrams show some numbers. What would you call these numbers? Why?

How could you work out the number without counting all the dots?

Some numbers can be represented *geometrically* using shapes made from dots. You have already done some work on square numbers and *triangular numbers*.

Square numbers can be represented using square arrays.

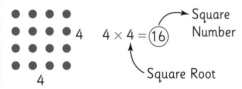

The number of dots on one side of the square is multiplied by itself to find the total number of dots. The total number of dots is the square number or *perfect square*. The length of the side is called the *square root*.

$4 \times 4 = \boxed{16}$ → Square Number

Square Root

Triangular numbers can be represented using dots to form triangles. The triangles can be isosceles or right-angled.

Both these sets of dots show the first four triangular numbers.

There is a difference of 1 between the number of dots in each row or column.

Each triangular number can be written as a sum of consecutive numbers.

1          $1 + 2 = 3$ $1 + 2 + 3 = 6$ $1 + 2 + 3 + 4 = 10$ and so on.

Rectangular numbers can be shown as rectangles with more than one row and column.

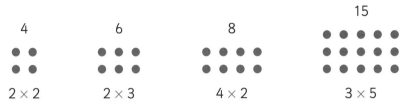

4	6	8	15
2×2	2×3	4×2	3×5

The rectangular number is the product of the length and width.

Rectangular numbers can be written as a product of two factors.

All rectangular numbers are composite numbers. They cannot be prime numbers but they can be square numbers.

Oblong numbers are a special type of rectangular number. These numbers can be shown as rectangles with a difference of 1 between the length and breadth.

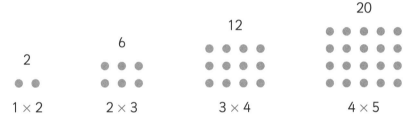

2	6	12	20
1×2	2×3	3×4	4×5

Oblong numbers can be written as the product of two consecutive whole numbers.

$0 \times 1 = 0$ $1 \times 2 = 2$ $2 \times 3 = 6$ $3 \times 4 = 12$ $4 \times 5 = 20$

There is one prime oblong number. Can you see it above?

1   Without drawing them, write down the 8th, 10th and 20th square numbers.

2   Look at this set of numbers.

3	5	7	8	12	30	17	11
36	15	100	40	19	25	21	12

   a   Which of these are square numbers?

   b   Find two triangular numbers in the set.

   c   Is 7 a rectangular number? Why?

   d   Draw a diagram to show that 30 is both rectangular and oblong.

   e   Which of the numbers are not square, triangular, rectangular nor oblong?

3   Micah built a number using counters. It is an oblong number. One side of the shape is 11 counters long. What could the other side be? Explain your answer.

4   Jessica says that 21 cannot be a rectangular number because it is a triangular number. Is she correct? Explain your answer.

5 Sharon is playing around with triangular numbers. She thinks that all triangular numbers can be rearranged to make rectangles. She tries with 15 and it works.

$1 + 2 + 3 + 4 + 5 = 15$ $3 \times 5 = 15$

a Does this method work for the next two triangular numbers: 21 and 28? Draw diagrams to check your answer.

b Sharon says that you can group the number into pairs and then multiply to find the answer faster when you are working with triangular numbers. She writes these notes in her book:

> $1 + 2 + 3 + 4 + 5 + 6 + 7 + 8 + 9 + 10$
>
> $= (1 + 10) + (2 + 9) + (3 + 8) + (4 + 7) + (5 + 6)$
>
> This is the same as $11 \times 5 = 55$
>
> If you work out the sum of each pair and how many there are,
>
> you do not need to write them all out.

Test Sharon's idea. Does it work?

6 Royston looks at the first five triangular numbers and he sees a pattern.

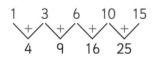

a Look at his notes and explain mathematically what he has seen.

b Do you think this will work for all triangular numbers? Give reasons for your answer.

Looking Back

1 Show that 6 is a triangular, rectangular and oblong number.

2 What is the fewest number of counters that you can move to show that this triangular number is also square? You can use real counters to help if you need to.

Unit 2 Use Patterns to Calculate

Let's Think …

Look at these three facts:

$4 \times 8 = 32$ $40 \times 8 = 320$ $400 \times 8 = 3\,200$

1 What pattern can you see?
2 How can this help you to do these two calculations mentally?
   $40\,000 \times 8$ $4 \times 8\,000$

A good understanding of place value and the skill of noticing and using patterns can help you find mental methods of calculating quickly.

You already know that patterns of multiplying and dividing by multiples of 10 are very useful.

$\times$	10	100	1000	10000	100000
12	120	1 200	12 000	120 000	1 200 000
23	230	2 300	23 000	230 000	2 300 000
145	1 450	14 500	145 000	1 450 000	14 500 000

When you divide, you may get decimal answers.

$\div$	10	100	1000
11 240	1 124	112.4	11.24
1 340	134	13.4	1.34
190	19	1.9	0.19

Making a list or a table can help you spot patterns and find methods of working out numbers further along the pattern.

1 Use the first number fact in each set and the patterns you know to find the answers to the other facts mentally.

  a $15 \times 7 = 105$, work out:
    150×7 15×70 15×700 150×700 150×70

  b $19 \times 8 = 152$, work out:
    190×8 $1\,900 \times 8$ 190×80 $19 \times 8\,000$ 190×80

  c $32 \times 6 = 192$, work out:
    320×6 $3\,200 \times 6$ $32 \times 6\,000$ $32 \times 60\,000$ 320×60

2 What methods and patterns did you use in question 1? Share your methods with your group.

3 Use the given division facts to help you work out a pattern and the answers to the other facts.

a	b	c
$9 \div 3 = 3$	$18 \div 2 = 9$	$38 \div 2 = 19$
$90 \div 30 = 3$	$180 \div 2 = 90$	$380 \div 20 = 19$
$90 \div 3 = 30$	$180 \div 20 = 9$	$380 \div 2 = ?$
$900 \div 30 = ?$	$1\,800 \div 2 = ?$	$3\,800 \div 2 = ?$
$9\,000 \div 30 = ?$	$18\,000 \div 2 = ?$	$3\,800 \div 20 = ?$
$900 \div 3 = ?$	$1\,800 \div 20 = ?$	$38\,000 \div 20 = ?$
$9\,000 \div 300 = ?$	$18\,000 \div 20 = ?$	$38 \div 20 = ?$

4 Work with a partner to try to find a pattern that will help you answer each of these calculations mentally.
   a Add 100 to any number.
   b Subtract 100 from any number.
   c Find half of a number.
   d Increase a number by 10.
   e Divide any number by 10.
   f Multiply any number by 100.

5 Kenya has drawn up this table to show the cost of her bus tickets for 4 weeks:

Number of Weeks	Cost of Bus Tickets ($)
1	12.00
2	24.00
3	36.00
4	48.00
5	?

She says you can work out costs by counting in 1s for the tens column and in 2s for the ones column.
   a Work out the cost for 5 weeks.
   b Is Kenya's method correct? Explain.
   c Can you find a rule for working out the cost for any number (n) of weeks?
   d Use your rule to work out the cost for 12 weeks.
   e After how many weeks will Kenya have spent $100.00 on bus tickets? How did you work this out mentally?

6 Ten thousand people attended an Olympic event. The tickets cost $25.50 per person.
   a How much money was spent on tickets?
   b The money raised at the event was divided evenly between 10 sporting charities. How much did each charity receive?

Looking Back
Choose one of the calculations from question 5. How would you teach someone to use the pattern or method you found?

Topic Review

What Did You Learn?

- A pattern is an ordered set of numbers or objects.
- A geometric pattern is made with shapes or objects. You can represent some numbers using shape patterns made from dots.
- A number multiplied by itself produces a square number.
- Numbers that can be shown as triangular arrays are called triangular numbers.
- Numbers that can be shown as rectangular arrays with more than one row and column are called rectangular numbers.
- Oblong numbers are formed by multiplying two consecutive numbers. They can be shown as rectangular arrays.
- Patterns can be used to make mental calculations more efficient.

Talking Mathematics

How can you use patterns to help convert between the following units in the metric system?

- Centimetres to millimetres
- Kilograms to grams
- Millilitres to litres

Prepare a short talk to teach someone how to do this.

Quick Check

1 Choose numbers from the box to answer the questions.

| 12 | 20 | 14 | 15 | 21 | 3 | 1 | 2 | 11 | 25 | 100 |

Which numbers are:

  a oblong　　　b square　　　c prime　　　d triangular　　　e oblong and rectangular?

2 Say whether these statements are TRUE or FALSE. Give a reason for your answer.

  a There are no prime square numbers.
  b All oblong numbers are also rectangular numbers.
  c A rectangular number can also be a square number.
  d All triangular numbers are also rectangular.
  e A square number can also be oblong.

3 Ten students pay the same amount for a school trip. The total amount is $36.50. How much did each student pay?

4 5.9 times an unknown number is equivalent to $5\,900 \div 100$. What is the unknown number?

Topic 4 Measurement Workbook pages 8–11

Key Words
length
volume
capacity
mass
temperature
time
Celsius (C)
Fahrenheit (F)
customary units

▲ Which measurement is an Olympic record for the 100 metres sprint:
9.58 seconds or 9.59 minutes? Would a cyclist cover the same distance
in a shorter or longer time?

Measurements tell you about how big or long something is, how hot or cold it is or how far away it
is. Sometimes you need to measure accurately and sometimes it is useful to be able to estimate a
measurement quickly.

In this topic, you are going to use what you already know and then learn more about measurements.

Getting Started

1 What do you measure if you need to know your height? What unit or units of measurement would you
use to do this?

2 Would you use kilograms or grams to measure the mass of a bag of potatoes? Why?

3 Which stick is longer: a stick that measures 1.2 m or a stick that measures 12 cm?

4 Can you measure the capacity of a triangle? Explain your answer.

5 Which of the following are metric units and which are customary units?

gallons   litres   feet   grams

6 Work with a partner to plan the route for a 500 metre long fun run at your school.
   ● Use measuring tapes or metre sticks to measure out the route.
   ● Draw a simple map of the route.
   ● How long do you think it will take a Grade 6 student to run this distance?

7 The mass of trucks and other large vehicles is often measured on a device called a weighbridge. How do
you think this works? What units of mass do you think it uses? Do some research to find out, if you do not
know the answers.

Unit 1 Use the Metric System

Let's Think …

1 What do you think the expression 'a rule of thumb' means? What does it have to do with measurement?
2 If you were building a new road, why would you need to take accurate measurements? What would you need to measure?
3 Which units of measurement would you use to make accurate measurements of the following:
   a the mass (weight) of a suitcase full of clothes
   b the amount of liquid in a medicine bottle
   c your body temperature
   d the height of a wall around your house.

Read through the table to revise the different units of measurement that you worked with in previous grades.

Measurements	Units of Measurement
length or distance	millimetre (mm), centimetre (cm), metre (m), kilometre (km)
volume	cubic units (e.g. cm³)
capacity	millilitre (mL), litre (L), kilolitre (kL)
mass	milligram (mg), gram (g), kilogram (kg), tonne (t)
temperature	Celsius (°C), Fahrenheit (°F)
time	second, minute, hour, day, week, month, year

1 Work in groups. Your teacher will give you a measurement topic. Discuss and list five different situations in which you would use this measurement. State the unit of measurement you would use in each situation.

2 Work in pairs or groups. Your teacher will ask you to make some real measurements of length or distance.
   ● Decide which unit of measurement is best for each measurement.
   ● Estimate each measurement and then take accurate measurements.
   ● Record your results in a chart like this:

What I Am Measuring	Unit of Measurement	Estimate	Accurate Measurement in Metric Units

3 Write down three things that you could weigh in each of these units:
   a milligrams   b grams   c kilograms.

4 Choose the unit that is the best choice to measure each of these items.
   litres (L), millilitres (mL) or kilolitres (kL)
   a The amount of cola in a can.   b The amount of water in a fish tank.
   c The amount of gas needed to fill a tank in a truck.   d The amount of crude oil in a tanker ship.

5 Put the following items in order from the smallest to the largest using mass and distance.

a **Mass**

- A small car
- A bag of sweet potatoes
- A laptop computer

b **Distance**

- The distance a car can drive in 5 minutes
- The distance from the halfway line on a football field to the goals
- The distance a person can walk in 10 minutes

You can measure temperature in degrees Celsius or degrees Fahrenheit.
$0\,°C = 32\,°F$

---

6 Which temperature matches each situation the best?

a The temperature of water from a hot water tap.

42 °C   20 °C   36 °F

b The temperature of water from a cold water tap.

55 °C   22 °C   12 °F

c Your normal body temperature.

36 °C   96 °C   55 °F

d The temperature of ice.

−15 °C   15 °C   150 °F

7 Write each of the following times in a different way.

a 7:00 p.m.   b 08:00   c four thirty in the afternoon

d midday   e half past two in the morning   f five minutes after midnight

8 You have to send some documents by post. The Post Office charges by the size and weight of each item you post. If your letter is 220 mm × 320 mm and it weighs 500 grams, how much will you pay?

Postage costs: Priority mail		
Size	**Weight**	
220 × 110 mm	first 100 g	$ 1.60
	each additional 50 g	$ 0.70
229 × 324 mm	first 100 g	$ 2.20
	each additional 50 g	$ 0.80

9 The temperature in a town in the Caribbean measured 34 degrees Celsius at midday. By six in the evening the temperature had cooled down by 4 degrees and by 23:00 the temperature had cooled down by a further 3 degrees. What was the temperature at eleven o'clock that evening?

Looking Back

Terri's uncle starts work at 08:15 every day. He works until 16:45. He has an hour's break for lunch each day and he works 5 days a week. How many hours does he work in one week?

Unit 2 Converting Units of Measurement

Let's Think ...

1 How much water should you drink each day: 2 *litres* or 2 *millilitres?* Explain why.
2 Which of these units could you use to measure the length or height of a table?
   feet   inches   metres   quarts   millimetres
3 Is this statement TRUE or FALSE?
   To change smaller units to larger units of measurement, you need to divide.

Converting Metric Units

To convert metric units you divide or multiply by 10 or multiples of 10 (such as 100, 1 000).

To convert millilitres to centilitres you multiply by 10.

To convert milligrams to grams you multiply by 1 000.

To convert metres to centimetres you divide by 100.

Decimal Notation

You can convert units in decimal notation by moving the digits left or right using place value. The numerals stay the same.

> 1.000 km = 1 000 metres
> 1.700 km = 1 700 metres (1.7 × 1 000)
> 2 300 mL = 2.300 litres (2 300 ÷ 1 000)
> 4 000 g = 4.000 kg (4 000 ÷ 1 000)

Customary Units

Customary units are not metric. They include units such as gallons, quarts, pints, inches and feet.

1 quart (qt) = 2 pints (pt)   1 gallon (gal) = 4 quarts (qt)
1 foot (ft) = 12 inches (in)   3 feet = 1 yard (y)

1 Study the step diagram. Use the diagram
   to convert the following units.

   a 25 decilitres = ___ litres
   b 3 kilograms = ___ grams
   c 55 millimetres = ___ centilitres
   d 250 centigrams = ___ grams

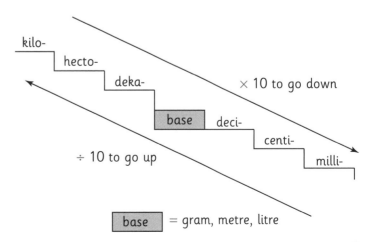

2   Work in pairs. Study the diagram below which shows how to convert metric units for distance and length. Explain to your partner how this works. Then make up questions to ask each other; for example: *An animal is 27 metres long. How many centimetres is that?*

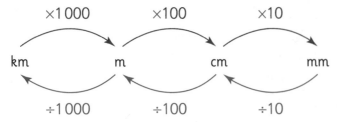

3   Convert the following measurements.

a 35 kg = ___ g

b 560 g = ___ kg

c ___ g = .450 kg

d 25 g = ___ kg

e ___ mL = 25 L

f 3 200 mL = ___ L

g 0.375 L = ___ mL

h 1.5 L = ___ mL

i 1.5 km = ___ m

j 2 500 m = ___ km

k 35 mm = ___ cm

l ___ cm = 3.2 m

4   In a science lesson, you follow instructions to make two solutions. You put 1 500 milligrams of salt in one 100 mL jug of water and 150 grams of salt in another 100 mL jug of water. Which solution has more salt?

5   Joni bought some pieces of ribbon in a sale. Put the lengths of ribbon in order from the shortest piece to the longest piece.

1.15 m   350 cm   2 500 mm   1.20 m   0.5 m

6   Work in pairs and discuss how to complete these statements.

a To convert pints to quarts, ___ by ___.

b To convert inches to feet, ___ by ___.

c To covert quarts to gallons, ___ by ___.

7   Convert these measurements.

a 5 foot 3 inches = ___ inches

b 6 pints = ___ gallons

c 8 quarts = ___ pints

d 24 inches = ___ feet

e 2 yards = ___ inches

Looking Back

Work in groups. Create 10 story problems about measurement. Make sure you know the correct answer. Then have a quiz, using the questions you have created. Discuss the rules of the quiz before you begin including how you are going to score the answers.

Topic Review

Talking Mathematics

What is the mathematical word for each of these?

- The amount a container can hold.
- The amount of space a solid figure occupies.
- The amount of matter in an object.
- A metric unit of length equivalent to 1 000 mm.
- Customary units used to measure capacity.
- A customary unit that equals 12 inches.

Quick Check

1   Which units of measurement would you use to measure the following?

   a   The length of a road.
   b   The capacity of a glass.
   c   The temperature in the classroom.
   d   The time it takes to get dressed in the morning.

2   Copy and complete this table of metric units of length.

	hecto-	deka-	metre			
1 000 m		10 m		0.1 m		0.001 m

3   Rewrite each set of measurements in ascending order.

   a   32 cm   0.025 cm   1.8 cm   330 mm   0.3 m
   b   150 cm   2.2 m   37.8 cm   1 300 mm   0.99 m

4   The wheel of Micah's bicycle travels 1.35 m each time it turns. Work out how far he cycles if the wheel turns:

   a   100 times   b   1 000 times

5   A bucket holds 3.4 litres and a full beaker of water holds 100 mL. How many beakers of water would you need to fill the bucket?

Topic 5 Fractions Workbook pages 12–13

▲ Our flag is one of the few in the world that is half as wide as it is long. What does this mean for a flag that is 1.2 metres long? Look at the colours of the flag. Do you think the blue takes up more or less than half the flag? Why?

You can find examples of **fractions** all around you. Think about the two halves of a sports field, equal sized pieces of cake and even the panes of glass in a window. In this topic, you are going to revise **equivalent** fractions and use common factors to **simplify** fractions. You are also going to **order** and **compare** fractions.

Getting Started

1 Our flag is approximately $\frac{2}{10}$ black and $\frac{2}{10}$ yellow. Estimate the fraction of the flag that is blue. Give your answer in simplest terms.

2 A school flag is $\frac{1}{4}$ blue, $\frac{2}{8}$ green and $\frac{3}{12}$ red. The rest is white. What fraction is white?

3 Look at these two diagrams.

   a What does the circle show in each case?
   b What fraction of each diagram is shaded yellow?
   c Describe the other colour shadings using fractions.
   d Which is greater, $\frac{2}{3}$ or $\frac{3}{10}$?
   e Which is smaller, $\frac{7}{20}$ or $\frac{2}{3}$?

Unit 1 Equivalent Fractions

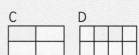

Let's Think ...

1 What fraction of each shape has been shaded yellow?
2 Which diagrams show equivalent fractions?
3 Write down the equivalent fractions.
4 Compare the yellow fraction of Shape A and Shape B. Which is greater?

A   B

C   D

A fraction is a number of parts of a whole.

The denominator tells you how many equal parts the whole is divided into.

The numerator tells you how many of the parts you are working with.

numerator
denominator $\dfrac{3}{4}$ 3 out of 4 equal parts

Mixed numbers have a whole number part and a fraction part.

They can be regrouped to have a numerator that is greater than a denominator to make them easier to work with.

Equivalent fractions have the same value.

Equivalent fractions can be found by multiplying or dividing the numerator and denominator by the same number.

mixed number

$3\dfrac{1}{7}$

whole number   fraction

1 whole $= \dfrac{7}{7}$

3 wholes $= \dfrac{7}{7} + \dfrac{7}{7} + \dfrac{7}{7} = \dfrac{21}{7}$

So $3\dfrac{1}{7} = \dfrac{21}{7} + \dfrac{1}{7} = \dfrac{22}{7}$

$\dfrac{1}{4} \times \dfrac{3}{3} = \dfrac{3}{12}$, so $\dfrac{1}{4} = \dfrac{3}{12}$

$\dfrac{7}{9} \times \dfrac{4}{4} = \dfrac{28}{36}$, so $\dfrac{7}{9} = \dfrac{28}{36}$

$\dfrac{6}{18} \div \dfrac{6}{6} = \dfrac{1}{3}$, so $\dfrac{6}{18} = \dfrac{1}{3}$

$\dfrac{30}{50} \div \dfrac{10}{10} = \dfrac{3}{5}$, so $\dfrac{30}{50} = \dfrac{3}{5}$

$\dfrac{1}{3}$ *is the simplest form of* $\dfrac{6}{18}$ *and* $\dfrac{3}{5}$ *is the simplest form of* $\dfrac{30}{50}$.

A fraction is in its simplest form when no whole number except 1 can divide into both the numerator and denominator.

To find the simplest form of any fraction you divide the numerator and denominator by the highest common factor of both numbers. There are different ways to do this.

Write $\dfrac{16}{40}$ *in its simplest form.*

Method 1

$\dfrac{16}{40} \div \dfrac{8}{8} = \dfrac{2}{5}$

Divide by the GCF = 8

Method 2

$\dfrac{16}{40} \div \dfrac{4}{4} = \dfrac{4}{10}$

$\dfrac{4}{10} \div \dfrac{2}{2} = \dfrac{2}{5}$

Work in steps

Method 3

$\dfrac{\overset{2}{\cancel{16}}}{\underset{5}{\cancel{40}}} = \dfrac{2}{5}$

Cancelling (a short way of writing the division)

1 Are these fractions equivalent?

 a $\frac{8}{12}$ and $\frac{2}{3}$ b $\frac{3}{10}$ and $\frac{3}{5}$ c $\frac{4}{12}$ and $\frac{1}{4}$

2 Divide the numerator and denominator of each fraction by the same number and write the fraction in simplest terms.

 a $\frac{56}{84}$ b $\frac{33}{99}$ c $\frac{42}{48}$ d $\frac{75}{100}$ e $\frac{84}{96}$

3 Write each fraction as an equivalent fraction with a denominator of 10.

 a $\frac{1}{2}$ b $\frac{6}{20}$ c $\frac{12}{30}$ d $\frac{4}{40}$ e $\frac{1}{5}$

4 Write each fraction as an equivalent fraction with a denominator of 100.

 a $\frac{1}{2}$ b $\frac{1}{4}$ c $\frac{3}{4}$ d $\frac{1}{5}$ e $\frac{1}{10}$

 f $\frac{3}{10}$ g $\frac{1}{20}$ h $\frac{1}{25}$ i $\frac{12}{50}$ j $\frac{25}{50}$

5 Write each fraction in simplest form.

 a $\frac{12}{32}$ b $\frac{15}{45}$ c $\frac{42}{49}$ d $\frac{36}{42}$ e $\frac{80}{100}$

 f $\frac{6}{10}$ g $\frac{27}{72}$ h $\frac{11}{33}$ i $\frac{18}{60}$ j $\frac{35}{45}$

6 Find and write down all the sets of equivalent fractions in the box. Circle the fraction in each set that is in simplest form.

$\frac{1}{5}$	$\frac{10}{16}$	$\frac{75}{100}$	$\frac{2}{7}$	$\frac{10}{50}$	$\frac{15}{24}$	$\frac{3}{4}$
$\frac{10}{14}$	$\frac{6}{30}$	$\frac{5}{8}$	$\frac{6}{8}$	$\frac{12}{16}$	$\frac{3}{15}$	$\frac{25}{400}$

7 How many equivalent fractions with a smaller denominator can you make for each fraction?

 a $\frac{160}{200}$ b $\frac{42}{63}$ c $\frac{42}{56}$

8 A fraction is equivalent to $\frac{2}{3}$ and the sum of its numerator and denominator is 15. What is the fraction?

Looking Back

Four fractions are equivalent in each set. Find the one that is not equivalent and write it in simplest terms.

Set A: $\frac{9}{21}$ $\frac{18}{42}$ $\frac{15}{27}$ $\frac{3}{7}$ $\frac{6}{14}$

Set B: $\frac{42}{60}$ $\frac{56}{70}$ $\frac{21}{30}$ $\frac{14}{20}$ $\frac{7}{10}$

Unit 2 Compare and Order Fractions

Let's Think …

Jayden and Alex have the same amount of homework. Jayden has completed $\frac{11}{15}$ of his homework and Alex has completed $\frac{4}{5}$ of his.

Who has done the most homework?

You can use the signs <, = or > to compare two fractions.

To compare fractions with the same denominators, look at the numerators.

$$\frac{1}{8} < \frac{5}{8} \qquad \frac{7}{9} > \frac{5}{9} \qquad \frac{17}{40} < \frac{30}{40}$$

To compare fractions with different denominators, use equivalent fractions.

Compare $\frac{2}{3}$ and $\frac{3}{5}$

$\frac{2}{3} \times \frac{5}{5} = \frac{10}{15}$ Change both fractions to get fifteenths

$\frac{3}{5} \times \frac{3}{3} = \frac{9}{15}$

$\frac{10}{15} > \frac{9}{15}$,   so $\frac{2}{3} > \frac{3}{5}$

You can write sets of fractions in ascending or descending order.

To order sets of fractions, write equivalent fractions with the same denominator.

Arrange these fractions in ascending order.

$$\frac{3}{5} \qquad \frac{3}{4} \qquad \frac{7}{10} \qquad \frac{1}{2}$$

Look at the denominators: 5, 4, 10 and 2

All of these denominators are factors of 20.

Change the fractions to make equivalent twentieths.

$\frac{3}{5} \times \frac{4}{4} = \frac{12}{20}$ $\frac{3}{4} \times \frac{5}{5} = \frac{15}{20}$

$\frac{7}{10} \times \frac{2}{2} = \frac{14}{20}$ $\frac{1}{2} \times \frac{10}{10} = \frac{10}{20}$

The order is $\frac{10}{20}, \frac{12}{20}, \frac{14}{20}, \frac{15}{20}$

Now you can write the original fractions in ascending order.

$$\frac{1}{2}, \quad \frac{3}{5}, \quad \frac{7}{10}, \quad \frac{3}{4}$$

These methods work for mixed numbers as well. But remember to compare the whole numbers first – it does not matter what the fraction part is if the whole numbers are different.

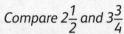

> Compare $2\frac{1}{2}$ and $3\frac{3}{4}$
>
> Look at the whole numbers: $3 > 2$, so $3\frac{3}{4} > 2\frac{1}{2}$

1 Which fraction is greater in each pair?

   a $\frac{2}{7}$ or $\frac{3}{8}$ b $\frac{9}{10}$ or $\frac{6}{10}$ c $\frac{11}{100}$ or $\frac{17}{100}$ d $\frac{4}{5}$ or $\frac{9}{10}$ e $\frac{5}{6}$ or $\frac{11}{12}$ f $\frac{6}{11}$ or $\frac{22}{33}$

2 Rewrite each fraction with a denominator of 48 and then arrange the original fractions in ascending order.

   $\frac{5}{6}$ $\frac{1}{3}$ $\frac{11}{12}$ $\frac{19}{24}$ $\frac{3}{4}$ $\frac{5}{8}$ $\frac{9}{24}$ $\frac{21}{24}$

3 Arrange each set of fractions in descending order.

   a $\frac{2}{15}$ $\frac{3}{5}$ $\frac{1}{10}$ b $\frac{2}{3}$ $\frac{4}{9}$ $\frac{11}{27}$ c $\frac{1}{3}$ $\frac{3}{4}$ $\frac{3}{8}$ $\frac{5}{12}$

   d $\frac{5}{12}$ $\frac{3}{4}$ $\frac{1}{3}$ $\frac{5}{6}$ e $\frac{3}{10}$ $\frac{1}{2}$ $\frac{3}{4}$ $\frac{2}{5}$ $\frac{1}{4}$ f $\frac{1}{2}$ $\frac{1}{4}$ $\frac{3}{10}$ $\frac{31}{100}$ $\frac{2}{5}$

4 The number line shows the position of some fractions.

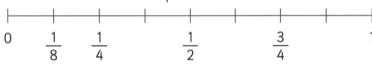

   0 $\frac{1}{8}$ $\frac{1}{4}$ $\frac{1}{2}$ $\frac{3}{4}$ 1

   Between which two of the values on the line would you place these fractions?

   a $\frac{3}{16}$ b $\frac{1}{10}$ c $\frac{7}{16}$ d $\frac{5}{8}$ e $\frac{7}{10}$ f $\frac{28}{32}$

5 Each sack weighed 1 kilogram when it was full. The mass of each one now is written on it as a fraction of a kilogram.

   a Write the masses in order from lightest to heaviest.   b Which sacks are more than half full?

 $\frac{3}{8}$ kg $\frac{4}{15}$ kg $\frac{3}{5}$ kg $\frac{1}{3}$ kg $\frac{7}{12}$ kg $\frac{2}{3}$ kg

6 The amount of homework that each student has completed is given below.

   | Jayden $\frac{5}{6}$ | Micah $\frac{7}{8}$ | Sharyn $\frac{6}{7}$ | Shayna $\frac{4}{5}$ | Dezi $\frac{9}{10}$ |

   a Write their names in order from most homework completed to least homework completed.

   b Which students have completed more than half their homework?

Looking Back

1 Which fraction is smaller in each pair?

   a $\frac{3}{4}$ or $\frac{2}{3}$ b $\frac{1}{4}$ or $\frac{5}{12}$ c $\frac{2}{3}$ or $\frac{11}{12}$

2 Write these fractions in ascending order.

   $\frac{1}{4}$ $\frac{6}{10}$ $\frac{2}{3}$ $\frac{1}{5}$ $\frac{5}{6}$ $\frac{7}{15}$

Topic Review

What Did You Learn?

- You write fractions using a numerator and denominator like this: $\frac{2}{3}$.
- The denominator shows how many parts the whole is divided into.
- The numerator tells you how many parts of the whole you are working with.
- A mixed number has a whole number part and a fraction part.
- Equivalent fractions have the same value.
- You can find equivalent fractions and simplify fractions if you multiply or divide the numerator and denominator by the same number.
- A fraction is in simplest form when the numerator and denominator cannot be divided by any number except 1.
- To order and compare fractions you can use equivalent fractions.

Talking Mathematics

Can you solve these riddles?

- I am a proper fraction. I am equivalent to $\frac{1}{5}$ but my denominator is 60. What am I?
- I am a fraction equivalent to $\frac{6}{8}$. My denominator is greater than 16 but less than 21. What am I?
- The numerator and denominator of my fraction add up to 35. In simplest form it would be $\frac{3}{4}$. What am I?

Quick Check

1 What is the simplest form of each fraction?

   a $\frac{6}{10}$ b $\frac{16}{24}$ c $\frac{20}{25}$

   d $\frac{15}{27}$ e $\frac{35}{100}$

2 One fraction in each set is not equivalent to the others. Which one is it?

   a $\frac{4}{18}$ $\frac{8}{36}$ $\frac{14}{63}$ $\frac{2}{9}$ $\frac{32}{108}$

   b $\frac{3}{4}$ $\frac{10}{12}$ $\frac{5}{6}$ $\frac{15}{18}$ $\frac{45}{54}$

3 Compare the fractions using the $<$, $=$ or $>$ signs.

   a $\frac{2}{8}$ and $\frac{1}{4}$ b $\frac{3}{4}$ and $\frac{7}{8}$ c $\frac{1}{5}$ and $\frac{2}{10}$

   d $\frac{3}{4}$ and 1   e $\frac{5}{10}$ and $\frac{1}{2}$ f $\frac{1}{3}$ and $\frac{4}{10}$

4 Write each set of fractions in descending order.

   a $\frac{7}{10}, \frac{3}{12}, \frac{1}{3}, \frac{1}{2}$ b $\frac{2}{8}, \frac{1}{3}, \frac{1}{2}, \frac{9}{12}$

   c $1\frac{2}{5}, 1\frac{3}{10}, 1\frac{6}{10}$ d $2\frac{1}{2}, 1\frac{1}{4}, 3\frac{5}{12}, 1\frac{1}{2}$

5 Write each set of fractions in ascending order.

   a $\frac{1}{2}, \frac{2}{3}, \frac{3}{4}, \frac{2}{6}$ b $1\frac{1}{8}, 1\frac{1}{5}, 1\frac{1}{12}, 1\frac{1}{2}$

6 Which is further, $\frac{65}{75}$ kilometres or $\frac{9}{10}$ of a kilometre?

7 I have a fraction card. The fraction on the card is equivalent to $\frac{1}{3}$. If I subtract 3 from the numerator and 6 from the denominator, the new fraction is half the value of the one I started with. What is the fraction on the card?

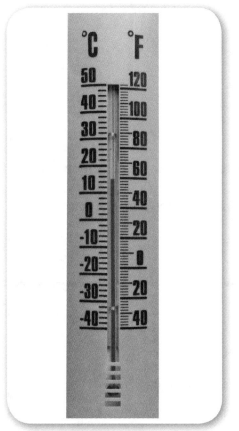

Key Words

integer

whole number

positive

negative

increase

decrease

◀ You already know how to read temperatures on a thermometer. At how many degrees Celsius does water freeze? What does it mean when we say that the temperature has fallen below zero degrees Celsius?

Whole numbers are the numbers starting from 0 that do not include fractions or decimals, for example the numbers we use when we count people. But what happens when we want to refer to whole numbers that are less than zero? In this topic, you will learn about the **integers**. Integers are the set of numbers that include all whole numbers, zero and their opposites.

Getting Started

1 Look at the diagram carefully.

  a Why do you think 0 is shown next to the water level?

  b Where is the bird in relation to the surface of the water? And the fish?

  c Which is further below the water level, the fish or the wreck?

2 How could you write the following in numbers?

  a 10 m above sea level

  b 10 m below sea level

  c 3 floors below ground level

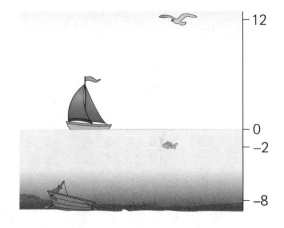

3 Where do you see or use negative numbers in daily life? Work in groups and list as many examples as you can; for example, to talk about temperatures below zero, on the buttons of an elevator and so on.

Unit 1 Positive and Negative Integers

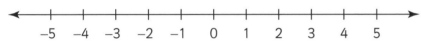

Let's Think ...

Look at the Celsius scale on the thermometer again.

- How is the scale like a number line?
- What happens to the numbers on the line as you go upwards?
- What happens to the numbers on the line as you go downwards?
- Why?

Whole numbers greater than 0 are called positive integers.

They are sometimes written with a positive sign (+), but if a number has no sign, it is always positive.

17   +45   +1 000   and   12 345   are all positive numbers.

Whole numbers less than 0 are called negative integers.

Negative integers are always written with a negative sign to show that they are negative.

−17   −45   −1 000   and   −12 345   are all negative numbers.

Zero (0) is neither negative nor positive.

The set of negative and positive integers (with 0) can be shown on a number line.

Each positive number has an opposite negative number.

```
←——+——+——+——+——+——+——+——+——+——+——+——→
  −5  −4  −3  −2  −1   0   1   2   3   4   5
```

You already know that positive numbers increase (get bigger) as you move to the right along the line and that they decrease (get smaller) as you move to the left along the line.

As you move left of 0, the numbers get smaller and smaller.

−1 means 1 less than 0, −2 means 2 less than 0 and so on.

You can use a number line to help you sort integers in order of size.

Look at the temperatures on the map. Write them in order from lowest to highest.

Draw a number line with positive and negative numbers. (You can stop at 4 as there are no numbers higher or lower than that).

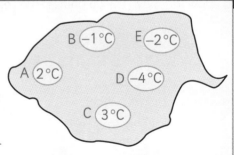

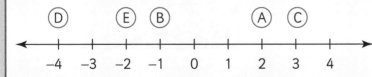

Mark and circle the temperatures on your number line.

The temperatures get higher from left to right, so the left-hand one is the lowest.

Write the temperatures in order from left to right.

−4°C, −2°C, −1°C, 2°C, 3°C

1 Write down the opposite of each integer.

   a 5   b −3   c 100   d 0   e −66

> Use this number line to help you in activities 2 to 6.
>
>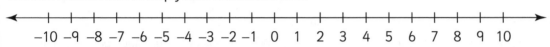
>
> −10 −9 −8 −7 −6 −5 −4 −3 −2 −1   0   1   2   3   4   5   6   7   8   9   10

2 Which is smaller in each pair?

   a 5 or −5   b −10 or 2   c −7 or −1

3 Write each set of numbers in descending order.

   a −8, 0, 2, −4, 5, −3, 8   b −3, 5, 4, 9, −9, −5, 1, 0

4 Follow the instructions, write the number you end on.

   a Start at −2 and move 5 jumps to the right.

   b Start at −2 and move 5 jumps to the left.

   c Start at 4, move 8 jumps left.

5 Write down all the integers between:

   a −3 and 7   b −5 and 1   c −6 and −2   d −9 and −2

6 Find a number that is:

   a even, greater than −6 and smaller than −3

   b odd, less than −4 and greater than −8

   c 8 more than −3

7 The heights above and below sea level are shown on this diagram.

   a At what height is the seagull?

   b What is at −2 m?

   c At what depth are the dolphins swimming?

   d How far above the dolphins is the seagull?

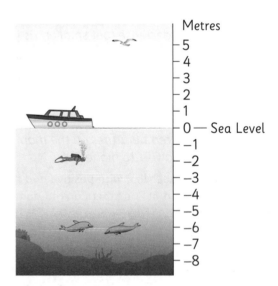

Looking Back

1 How many spaces from 0 is 4 on a number line? And how many from 0 to −4?

2 Is 0 a negative or a positive number?

3 Why is −15 smaller than −5?

Topic Review

What Did You Learn?

- An integer is a number with no fractional part. It can be positive, negative or zero.
- Positive integers are the numbers to the right of 0 on the number line. They can be written with or without a positive sign.
- Negative integers are the numbers to the left of 0 on the number line. They are always written with a negative sign.
- 0 is neither positive nor negative, but it is a member of the set of integers.

Talking Mathematics

Give the mathematical term for the following descriptions.

- Whole numbers that have a negative sign.
- All the whole numbers larger than 0.
- The set of all whole numbers and their opposites.

Quick Check

1 What are the missing words in these sentences?

  a On a horizontal number line, ___ numbers are to the left of zero.

  b A number with no sign is a ___ integer.

  c When you mark a positive integer on a horizontal number line, you would mark it on the ___ of zero.

2 Which of the following numbers is not an integer? Give a reason for your answer.

  6   1.5   −7   0

3 The temperatures (in °C) at 6 different stations are given in the table.

Place	A	B	C	D	E	F
Temperature (in °C)	−2	8	5	−4	0	−9

  a Which place has the highest temperature?

  b What is the lowest temperature?

  c Write the temperatures in ascending order.

4 The picture shows the buttons on an elevator in a hotel.

  a How many floors are there below ground level?

  b Mr Jones parks his car in the underground car park and goes up four floors. On which level does he get out?

  c Mrs Gladwell's room is on the fifth floor. How many levels will she go down to get to the gym?

  d Johnny is playing in the elevator. He gets in at the second floor. He goes up three levels, then down five, then up two and then down four. On what level is he now?

(5)	Fifth Floor
(4)	Fourth Floor
(3)	Third Floor
(2)	Second Floor
(1)	First Floor
(0)	Ground Floor & Reception
(−1)	Gym & Dining Room
(−2)	Underground Car Park

Topic 7 Decimals and Percentages Workbook pages 15–16

▲ Identify the decimals in the photograph. What does each one mean? Why do these decimals only go to hundredths? Why do some prices not have a decimal part?

Last year, you learned how to read, write, order and compare **decimals** up to **thousandths**. You know that each number after the **decimal point** represents a fraction. In this topic, you will revise decimal numbers up to three decimal places and you will compare and order decimals. You will learn about a new type of fraction, the **percentage**, and see how to express the same amount as a fraction, a decimal or a percentage.

Getting Started

1 The following numbers need decimal points. Decide where the decimal point should go to make each value realistic.

a	The top speed of a cheetah	1158 km/h
b	The price of a pair of running shoes	$4595
c	The average weight of a 12-year-old boy	399 kg
d	The test score of the top mathematics student	972%
e	The average height of a newborn baby	4955 cm

2 Make a list of places where you can see decimals used in daily life. Compare your list with a partner when you have finished.

3 Do you remember how to round decimals to the nearest whole number?

   a Write the steps you would follow to round 14.67 to the nearest whole number.

   b Give two examples of when a rounded decimal value is good enough to use.

   c Give two examples of where you would need to use the exact amount and not a rounded value.

Unit 1 Revisit Decimals

Let's Think ...

1 What is the digit in the hundredths place in the number 88.72?
2 What is the digit in the tenths place in the number 73.01?
3 For the number 59.138, what is the place value of the digit 8?
4 For the number 123.631, what is the value of the digit 6?

Decimals are numbers that use a decimal point and place value to show fractions. The number to the left of the decimal point is a whole number, while the numbers to the right show a number that is smaller than 1 and can be written as a fraction.

The number 43.725 consists of the following:

- *43 is a whole number*

- *7 tenths or $\dfrac{7}{10}$*

- *2 hundredths or $\dfrac{2}{100}$*

- *5 thousandths or $\dfrac{5}{1\,000}$*

Thousands	Hundreds	Tens	Units	•	Tenths	Hundredths	Thousandths
		4	3	.	7	2	5

*When you read this number, you say forty-three **and** seven hundred twenty-five thousandths.*

As a mixed number, it would look like this: $43\dfrac{725}{1\,000} = 43\dfrac{29}{40}$ in simplest form.

You can also write the number in expanded form, like this:

$$43.725 = 40 + 3 + \dfrac{7}{10} + \dfrac{2}{100} + \dfrac{5}{1\,000}$$

To compare decimals, you line up the decimal points and then compare each place value starting from the left.

Order the numbers 3.191, 3.172 and 3.177 from the least to the greatest.

3	.	1	9	1	The first two digits of each number are the same.
3	.	1	7	2	Look at the digits in the hundredth places. The first number is the largest so that is the greatest number.
3	.	1	7	7	The second and third numbers have the same hundredth values, so look at the thousandths place values. You can see that the second number is the least.

The order is 3.172, 3.177, 3.191

You can also compare decimals and fractions.

To do this, you convert the fraction to a decimal or the decimal to a fraction.

Use $<$, $=$ or $>$ to compare 0.125 and $\frac{1}{8}$.

Write the decimal as a fraction and simplify it. $0.125 = \frac{125}{1000}$

$$\frac{125}{1000} \div \frac{25}{25} = \frac{5}{40}$$

$$\frac{5}{40} \div \frac{5}{8} = \frac{1}{8}$$

Now you can see that $0.125 = \frac{1}{8}$

Always make sure that the values are in the same format when you want to compare them.

1 Write the following decimal numbers in expanded notation.

   a 518.46   b 3.729   c 23.233

2 Write each number as a decimal.

   a Three and fifty-eight hundredths

   b Two hundred forty-four and one tenth

   c Nine thousand seven hundred twenty-nine and eighty-five hundredths

   d Seventy-three and seven hundred twelve thousandths

   e Twenty-three hundredths

3 Use the symbols $<$ and $>$ to compare these decimals.

   a 23.07 ☐ 23.004   b 67.912 ☐ 67.92   c 7.74 ☐ 7.47

   d 1.099 ☐ 1.089   e 300.099 ☐ 300.01   f 3.009 ☐ 30.09

4 Four babies are born on the same day. Baby André weighs 3.375 kg, Baby Benni weighs 3.308 kg, Baby Colin weighs 3.37 kg and Baby Denae weighs 3 285 kg. Write the babies' names in order from the heaviest baby to the lightest baby.

5 In a javelin competition, each person gets six throws but only the best three results count. These are Keshorn's results for six throws.

   84.58 m   84.39 m   85.28 m   82.67 m   85.38 m   84.38 m

   a Arrange the results in order from least to greatest distance.

   b Which three results would be counted in this competition?

   c The Olympic record for men's javelin is 90.57 m (set in 2008). Look at Keshorn's best throw in this competition. How much further would he need to throw the javelin to beat the Olympic record?

   d In the next competition, Keshorn throws $\frac{3}{10}$ metre further than his best in this competition. What is that distance?

Looking Back

1 Work with a partner. Take turns to read the following numbers aloud:

   15.37   333.3   751.25   98.321   2.604

2 Write each set of decimals in ascending order.

   a 12.92, 13.58, 13.29, 13.51, 12.78

   b 0.23, 0.003, 0.203, 0.2, 0.3

Unit 2 Percentages

Let's Think ...

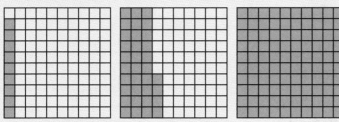

Look at each 10×10 grid above.

How many squares are shaded in each grid?

Express the shaded part of each grid as a fraction and as a decimal.

Per cent means 'for each hundred'. The symbol % is read as per cent and it shows that you are dealing with a *percentage*.

A percentage is really a fraction with a denominator of 100; for example, seventy per cent is $\frac{70}{100}$ or 70%.

One hundred per cent, 100%, is the whole. (Remember $\frac{100}{100} = 1$)

Look at the grids in the Let's Think ... again.

The first grid has 9 squares shaded out of the 100.

This can be written in three ways: $\frac{9}{100}$ 0.09   9%

These are different ways of writing the same value so they are *equivalent*.

$\frac{9}{100} = 0.09 = 9\%$.

To convert a percentage to a decimal, write it as a fraction with a denominator of 100, then convert the fraction to a decimal.

$85\% = \frac{85}{100} = 0.85$

Remember that a percentage is really a number of hundredths.

To convert a decimal to a percentage, write it as a fraction with a denominator of 100 and then as a percentage.

$0.72 = \frac{72}{100} = 72\%$

To write a fraction as a percentage, convert it to an equivalent fraction with a denominator of 100.

$\frac{2}{5} = \frac{2}{5} \times \frac{20}{20} = \frac{40}{100} = 40\%$

1   Write down five examples of where you would find percentages in daily life.

2   Julie has $100.00. She buys clothes costing $47.00 and spends $10.00 on lunch.
   a   What percentage of her money has she spent?
   b   What percentage of her money does she have left?

3   What percentage of each shape is shaded yellow?

   a   b   c

   d   e   f

   g   h   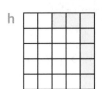   i

4   Convert each decimal to a percentage.
   a   0.45   b   0.7   c   0.35   d   0.25   e   0.07   f   0.3   g   0.01   h   0.5   i   0.8   j   0.99

5   Covert each fraction to a percentage.
   a $\frac{6}{10}$ b $\frac{9}{10}$ c $\frac{18}{100}$ d $\frac{8}{100}$ e $\frac{99}{100}$ f $\frac{1}{2}$ g $\frac{4}{5}$ h $\frac{9}{25}$ i $\frac{31}{20}$ j $\frac{5}{8}$

6   Write each percentage as an equivalent fraction and decimal.
   a   89%   b   52%   c   6%   d   100%   e   4%   f   90%   g   30%   h   16%   i   25%   j $33\frac{1}{3}$%

7   A rectangle is divided into 5 equal parts. What percentage of the rectangle is:
   a   one part   b   three parts   c   five parts

8   What percentage is left in each situation?
   a   Janae spent 15% of her pocket money on sweets and 32% on clothes.
   b   Leshawn spent 20% of his homework time on history and 34% on mathematics.

9   Anna scored $\frac{29}{40}$ in an English test and $\frac{17}{25}$ in a history test. In which subject did **she** get the highest percentage?

10 In Ms Walker's mathematics classes, 20 out of 100 students scored an A grade for a test. In Mr Darville's class, 4 out of every 20 students received an A grade. What percentage of each class scored an A grade?

11 Companies sometimes offer a percentage more food in a container at no extra cost. Look at the pictures and work out the mass of the special offer tin each case.

Looking Back

Prepare a short presentation on percentages, fractions and decimals.

Explain how percentages, fractions and decimals are equivalent, using examples.

Look in newspapers and magazines for examples of percentages, fractions and decimals, and explain when it is more suitable to use a percentage, a fraction or a decimal.

Topic Review

What Did You Learn?

- To compare and order decimals with the same whole number parts, use the first decimal place where the digits are different.
- Percent means parts per hundred. You indicate percentages with the % symbol.
- The same value can be expressed as a fraction, a decimal or a percentage. Since the terms have the same value, they are called equivalent.
- You can convert decimals to fractions by writing them with a denominator of 10, 100 or 1 000 and then simplifying them.
- You can convert fractions and decimals to percentages by making equivalent fractions with a denominator of 100.

Talking Mathematics

- How are decimals, fractions and percentage similar? How are they different?
- Use the advert to the right as an example to explain why the decimal point is important.

Large Lemonade
$100.00

Quick Check

1   What is the value of each of the digits in red?

   a 5525.37   b 517.139   c 316.02   d 55.193   e 300.003

2   Write each fraction as a decimal and as a percentage.

   a $\frac{7}{10}$ b $\frac{323}{1000}$ c $\frac{15}{10}$ d $\frac{9}{100}$ e $\frac{12}{25}$

3   Write down the remaining part of the pizza as a fraction, a decimal and a percentage.

4   For each diagram, write the percentage of one part of the shape and the percentage of the shape shaded blue.

   a 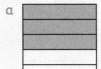   b   c   d   e

5   Which is greater in each pair?

   a 40 out of 160 or 30 out of 150?   b 20 out of 30 or 70 out of 100?

Topic 8 Classifying Shapes Workbook pages 17–22

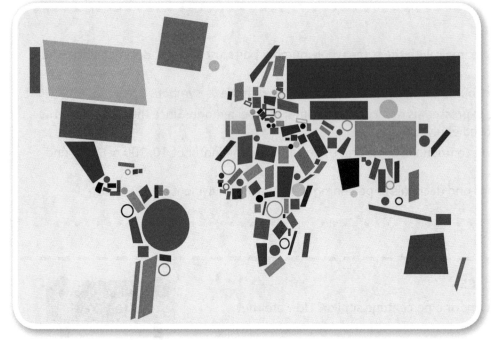

Key Words
acute angle
right angle
obtuse angle
straight angle
quadrilateral
equilateral
scalene
isosceles
centre
circumference
radius
diameter

▲ What does this picture show? How is it different from what you usually see on a world map? What shapes do you recognize in the picture above? What do you know about the shapes that you can identify?

Learning about shapes helps us to talk and think about the three-dimensional world that we live in. Some people use geometry at work every day, for example architects, carpenters and tailors. But even when we do things like wrap presents, build something or give directions to someone, we use skills that relate to shape, size and space. In this topic, you will revisit **angles** and **quadrilaterals**. You will also learn more about triangles and circles.

Getting Started

1 Which geometric shapes have angles? Are there any that do not have angles?

2 What is the smallest number of angles a geometric shape can have? And the largest number?

3 What happens to the size of angles in a regular polygon as you add more and more sides?

4 What do the following mean when applied to shapes?

   a tri-   b quad-   c hex-   d octa-

5 Read the riddles. Work out which shape it is.

> I am shaped like a box and all my faces are regular quadrilaterals.

> I have eight angles and I am often used for the red signs that tell drivers to stop.

> I have three sides. Two of them are the same length. I also have one angle measuring 90 degrees.

Unit 1 Angles

> **Let's Think …**
>
> - Explain what an angle is.
> - Name the different parts of an angle shown on the diagram.
> - Does the length of the arms affect the size of the angle? Explain.

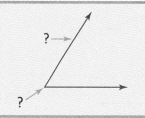

An angle is formed by two rays or line segments that meet at a point. The meeting point is called the vertex. The sides of the angle are called its arms.

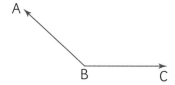

This angle is named ∠ABC, or ∠CBA. Since point B is the vertex, it must always be in the middle. The vertex can also be used on its own to refer to the angle: B̂.

Angles are classified by size.

An angle that measures between 0° and 90° is called an *acute angle*.

An angle that looks like one corner of a square and measures exactly 90° is a *right angle*.

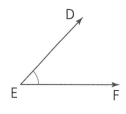

An angle that measures between 90° and 180° is called an *obtuse angle*.

An angle that measures exactly 180° and looks like a straight line is called a *straight angle*.

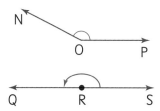

1 Three angles are marked on the diagram.
   a Name each angle using three letters.
   b Write the type of angle each is next to its name.

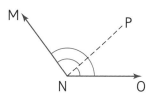

2 Without measuring, write down what type each of these angles is.

a

b

c

d

e

f

3 Measure each of the following angles with a protractor and write down its size.

a

b

c

d

e

f

4 Write down an estimate of the size of each of the angles below. Then measure the angle using a protractor and write down its actual size. Compare your answers with a partner to see whose estimates were the closest.

a

b

c

d

e

f

Looking Back
Make an informative poster using different materials such as string, straws or sticks to show the different parts of an angle as well as the different types of angles you have learned about.

Unit 2 Triangles

Let's Think ...

Draw any two triangles.
Measure the size of each triangle's three angles.
Add them together. What do you get?
Compare your answer to those of your classmates sitting near you.

A triangle is a polygon with three sides and three angles. The sum of the three angles of a triangle is always 180°.

You can name triangles according to how many sides are equal.

An *equilateral* triangle has three equal sides. All three angles are also equal and measure 60°.

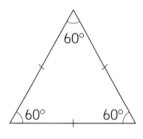

An *isosceles* triangle has two equal sides. The two angles at the base of the equal sides are also equal.

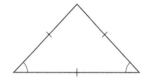

A *scalene* triangle has no equal sides and no equal angles.

Triangles can also be classified using the size of their angles.

An acute triangle has three acute angles.

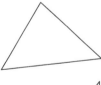

A right-angled triangle has one angle that measures 90°. The side opposite the right angle is always the longest side and is called the hypotenuse.

hypotenuse

An obtuse triangle has one obtuse angle.

1 Name each type of triangle as accurately as you can.

a    b   c   d   e

2 a What is the length of the third side in this equilateral triangle?

5cm

5cm

?

b What is the size of ∠ABC?

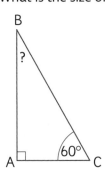

B

?

A   60°   C

c Which side of △KLM is the longest and which side is the shortest?

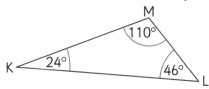

M

110°

K   24°   46°   L

d What is the length of QS?

Q

6 cm

R   S

e Write down any three possible measurements for the angles of an acute triangle.

3 The sizes of two angles of each triangle are given. Work out the size of the third angle and classify each triangle as acute, right-angled or obtuse.
   a 10°   40°   b 40°   50°   c 120°   30°   d 43°   38°

Looking Back

1 Why does each angle of an equilateral triangle have to measure 60°?
2 What is the sum of the other two angles of a right-angled triangle?
3 Can an obtuse triangle also be an isosceles triangle?
4 Is it possible to have more than one right angle in a triangle? Explain your answer.

Unit 3 Quadrilaterals

Let's Think ...

- 'Quad' means four and 'lateral' means side. So what does 'quadrilateral' mean?
- Find five examples of quadrilaterals around you. List them and draw a sketch of each.
- How are the quadrilaterals you found similar?
- How are they different?

A *quadrilateral* is any polygon with four sides and four angles.

A *square* is the only regular quadrilateral. It has four equal sides and four right angles.

Both pairs of opposite sides are parallel.

The small lines on each of the four sides show that they are equal. The small square in each angle indicates that it is a right angle. The parallel sides are marked with arrow pairs.

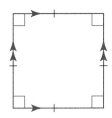

A *rectangle* also has four right angles. Its opposite sides are equal in length and parallel to each other.

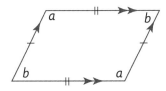

Parallelograms have two pairs of equal and parallel sides. However, its angles do not have to be right angles. Opposite angles are equal.

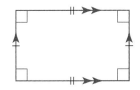

A *rhombus* has four equal sides with opposite sides parallel to each other and opposite angles equal.

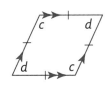

A *trapezium* (sometimes also called a trapezoid) has only one pair of parallel sides.

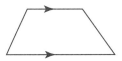

1 Name the following quadrilaterals.

a    b    c 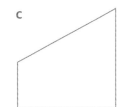   d    e

2 What do the quadrilaterals in each group have in common?

a

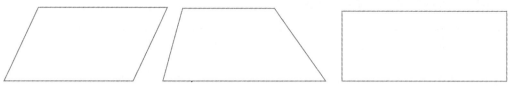

b

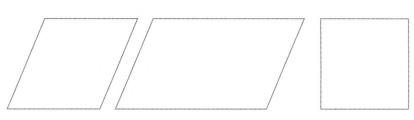

c

3 Measure all four angles of each quadrilateral below and add them together. What did you discover?

4 Write the names of quadrilaterals that have:
- **a** two pairs of opposite sides parallel
- **b** four sides of the same length
- **c** opposite angles that are equal in size
- **d** four right angles.

Looking Back

Complete the second Quadrilaterals activity in your Workbook.
Then draw a diagram that shows the relationship between a parallelogram, a rectangle and a square.

Unit 4 Circles

A circle is a closed plane shape with all points on its edge the same distance from the centre.

The distance around the circle is called the circumference of the circle.

The radius of a circle is a line segment that runs from the centre of the circle to any point on the circumference of the circle. The radius will always be the same length.

The diameter of a circle is the line segment that passes from a point on the edge of the circle through the centre and to the opposite edge. The diameter divides a circle into two equal halves. The length of the diameter is twice the length of the radius. The diameter of a circle will always be the same length.

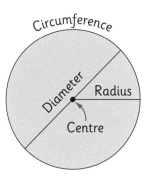

1 Name the parts of the circle shown below.

a  b

c 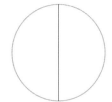 d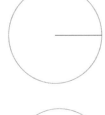

2 The radius of a circle is 2.5 cm long. How long is the diameter?

3 The centre point of a circle is 11 cm from its edge. What is the radius of the circle?

4 Steve runs a circular route three times a week. Today, he did five rounds and 2.5 km. What is the circumference of the circle?

5 Measure and identify the type of angle formed by the two radii of the circle.

a 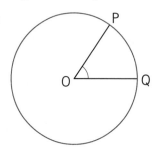 b

c 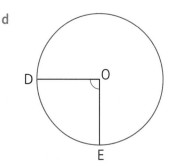 d

6 Micah cut a slice from a circular pizza. The pizza has a diameter of 25 cm. Work out the length of each straight side of the slice.

Looking Back

Draw a circle and use different colours to indicate the centre, circumference, radius and diameter. Measure the radius and then measure the diameter. Is your circle's diameter twice the length of the radius?

Topic Review

What Did You Learn?

- An acute angle measures between 0° and 90°. A right angle measures exactly 90°. An obtuse angle measures between 90° and 180°. A straight angle measures 180°.

- A quadrilateral is a polygon with four sides and four angles. Squares, rectangles, parallelograms, rhombuses and trapeziums are quadrilaterals with special properties.

- Triangles are named according to their sides and angles. Equilateral triangles have three equal sides and three equal angles, isosceles triangles have two equal sides and two equal angles.

- A circle is a shape that is made up of all the points on a plane that are the same distance from a given point. This point is called the centre of the circle.

- The circumference of a circle is the distance around the circle. The radius is the distance from the centre to the edge of the circle. The diameter is a line segment that goes across the circle through the centre. The diameter is twice the length of the radius.

Talking Mathematics

Match the mathematical term on the left with its best description on the right.

Trapezium	A line that goes from one edge of a circle through its centre to the other edge.
Equilateral triangle	The distance around a circle.
Acute angle	A triangle with three equal sides and three equal angles.
Diameter	A triangle with no equal sides or angles.
Obtuse triangle	A quadrilateral with one pair of parallel sides.
Circumference	A quadrilateral with four equal sides, opposite sides parallel and opposite angles equal.
Radius	A triangle with one obtuse angle.
Scalene triangle	A triangle with two equal sides and two equal angles.
Rhombus	A line segment from the centre to the circumference of a circle.
Isosceles triangle	An angle between 0° and 90°.

Quick Check

1. What do you call an angle that is bigger than a right angle but smaller than a straight angle?

2. Which quadrilaterals have four right angles?

3. Is a trapezium a special type of parallelogram?

4. Can an acute triangle be scalene, isosceles and equilateral?

5. Can you work out the length of the radius of a circle if you have the length of its diameter?

6. Identify as many different shapes as you can in the diagram. Use letters to name each shape you find and write the type of shape next to each one.

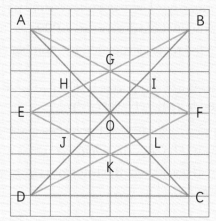

Topic 9 Rounding and Estimating Workbook pages 23–24

▲ The exact masses of these four sweet potatoes are 369 g, 378 g, 399 g and 382 g. Is it fair to say they each have a mass of about 400 g? Why? Can you think of a quick method of estimating the total mass of eight similar sweet potatoes? Share your ideas with your group.

Rounded or **approximate** numbers are often used in everyday life when exact values are not important. People often say things like: 'It will take about 2 hours to get to the beach.', 'I need about 300 bricks to pave my driveway.' or 'Nearly 5 000 people are expected to attend the cricket match on Saturday.' These are approximate numbers which are close enough to the exact values to make sense. Rounding and estimating are very useful for calculating as well. You are going to practice these skills so that you can use them effectively.

Getting Started

1 A news reporter says 'There are about 4 000 people watching the match.'

   a Does this mean there are exactly 4 000 spectators? Explain why or why not.

   b Assuming the reporter rounded the number correctly, could there be more than 4 000 spectators? Why?

   c Would the match organizers be likely to work with approximate figures to work out how much money they raised from ticket sales? Why?

2 Mr Samuels weighs some sweet potatoes and finds they have an average mass of 0.345 kilograms.

   a Is he correct if he says the average mass is about 0.5 kilograms? Why?

   b What would you expect the total mass of five of these sweet potatoes to be? How did you calculate your answer?

Unit 1 Revisit Rounding

Let's Think ...

Estimate the total length of these lines in centimetres without doing any measurements.

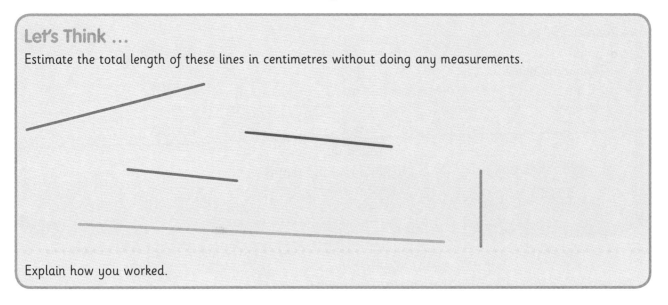

Explain how you worked.

The rules for *rounding* whole numbers and decimals are the same.
- *Find the* digit *in the rounding* place.
- *Look at the digit to the right of this place.*
- *If the digit to the right is 0, 1, 2, 3, or 4, leave the digit in the rounding place as it is.*
- *If the digit to the right is 5, 6, 7, 8, or 9, add 1 to the digit in the rounding place.*
- *Change all the digits to the right of the rounding place to 0.*

Example 1

Round 3 812 345 876 to the *nearest*:

a *hundred thousand*   b *ten million*   c *billion*

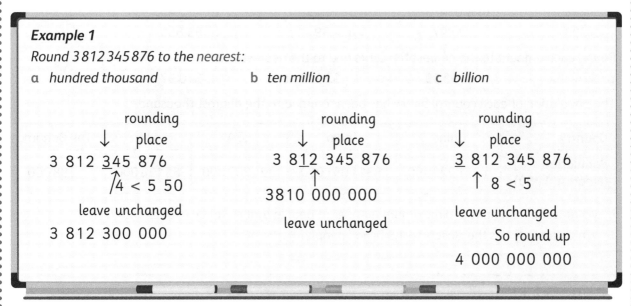

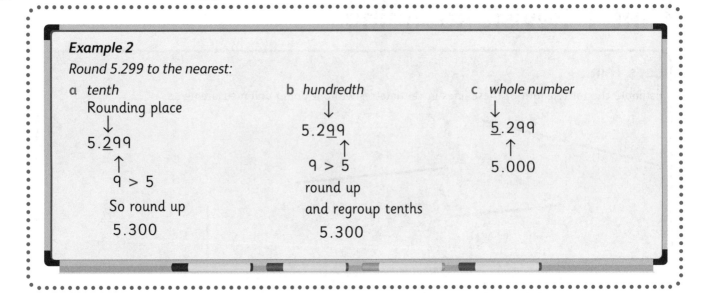

Example 2

Round 5.299 to the nearest:

a tenth
Rounding place
↓
5.2̲99
↑
9 > 5
So round up
5.300

b hundredth
↓
5.29̲9
↑
9 > 5
round up
and regroup tenths
5.300

c whole number
↓
5̲.299
↑
5.000

1 Round these numbers to make them more user-friendly.
   a There are 192 399 students in our school district.
   b Last year, 25 456 876 tourists visited the Caribbean.
   c The population of the world in November 2016 was 7 457 653 538.
   d Tourism contributes $48 987 654 345.67 to the Caribbean economy each year.
   e I live 9.99 km from school.
   f A bakery uses 99.826 kilograms of flour every week.

2 Round each decimal to the nearest hundredth, tenth and whole number. Write the three rounded figures next to each other.
   a 36.995   b 0.654   c 10.099   d 172.453
   e 5.298   f 12.997   g 1.99   h 65.855

3 Round each amount to the nearest ten cents and to the nearest dollar.
   a $144.82   b $198.99   c $2 999.46   d $10.97

4 The population of each country below has been rounded to the nearest thousand.

Country	China	India	USA	UK	The Bahamas
Population (Nearest 1 000)	1 371 220 000	1 311 051 000	324 709 000	65 138 000	388 000

   a What is the minimum and maximum number of people that could be in each country?
   b Which country has the closest to 1.5 billion people?

Looking Back
Use the figures from the table above.
a Round the populations of China, India, the USA and the UK to the nearest million.
b Round the population of The Bahamas to the nearest hundred thousand.

Unit 2 Estimate Answers

> **Let's Think ...**
>
> 12308 tourists visited a market one week and 12255 visited the next week. Discuss how you could estimate the total number of tourists by:
> - adding
> - multiplying.

Estimating is a very useful strategy to help you calculate quickly and to help you decide whether your answer is reasonable or not.

You can use rounding to estimate an approximate answer.

When you are working with whole numbers, you can usually round each number to its greatest place value. This is called leading figure rounding.

Example 1

Estimate 468×62

500×60 Round each number to the first digit (greatest place value).

$5 \times 6 = 30$ Use the facts you already know.

so $500 \times 60 = 30000$

$468 \times 62 \approx 30000$ $\approx$ means 'approximately equal to'

When you give an estimated answer, you use the $\approx$ symbol.

In some cases you do not need to round both numbers in a calculation.

Example 2

Estimate 12345×99

$12345 \times 100 = 1234500$ It is easy to multiply by 100, so only round the 99.

So, $12345 \times 99 \approx 1234500$

When the numbers you are working with are all similar in value, you can use clustering to estimate the answer.

Example 3

Estimate $416 + 438 + 447 + 407$

The numbers are clustered close to 400.

$400 \times 4 = 1600$

So, $416 + 438 + 447 + 407 \approx 1600$

1 Estimate each of the following. Choose the most efficient method.

  a 1 239 + 1 442
  b 3 229 + 11 187
  c 234 + 223 + 247 + 230
  d 132 876 + 132 087
  e 1 499 − 367
  f 21 148 − 349 + 424
  g 3 256 × 12
  h 589 × 52
  i 132 876 − 112 9 09
  j 3.63 + 3.8 + 3.21
  k 18.213 − 10.175
  l 230.464 − 64.593

2 A pilot flew 323.47 km on Monday, 338.43 km on Tuesday, 319.56 km on Thursday and 323.2 km on Friday. Estimate how far she flew in total.

3 Estimate the total cost of the following items in whole dollars.

  a $28.80 + $49.99 + $49.50 + $50.00
  b $11.82 + $12.20 + $12.09 + $12.12
  c $19.99 + $19.99 + $19.90 + $19.50

4 The numbers of tourists passing through a busy airport during a week are given in the table.

Day	Monday	Tuesday	Wednesday	Thursday	Friday
Number of Tourists	113 689	112 908	113 579	112 798	113 209

  a Use leading figure rounding to estimate the total number of tourists over the five days.
  b Use clustering to estimate the total.
  c Compare the two estimates. Which one is a more reliable figure? Why?

5 There are 86 400 seconds in a day. Estimate how many seconds there are in:

  a five days
  b four weeks.

6 Crowd attendance at a cricket match over a three-day period was 14 146, 15 964 and 17 193.

  a Estimate the total attendance.
  b If tickets cost $9.50 per day, estimate how much money was spent on tickets over the three days.

Looking Back

The prices of six homes that are for sale are given below.

If the estate agent manages to sell all six houses for close to the given price, estimate how much money will be paid in total.

Explain to your partner how you worked out your answer.

Topic Review

What Did You Learn?

- You can round numbers and decimals to any place using the same rules.
- If the digit to the right of the rounding place is 5 or greater, add one to the digit in the rounding place.
- If the digit to the right of the rounding place is 4 or less, leave the digit in the rounding place unchanged.
- Replace the digits to the right of the rounding place with zeros.
- Rounding is useful for estimating when you do not need an exact value.
- You can round whole numbers to their largest place value or leading figure.
- Sometimes it makes sense to only round one figure in the calculation.
- When the values are similar, you can cluster them and multiply to estimate the answer.
- When you have estimated an answer, you use the ≈ symbol. This means 'approximately equal to'.

Talking Mathematics

1 Give three examples of where you might use or hear rounded numbers in your everyday life.
2 When is it important to work with exact values and not rounded figures or estimates? Why?
3 How can clustering be useful when you are shopping?

Quick Check

1 Round to the nearest million.
   a 12 317 989   b 2 345 098 765   c 945 876   d 146 909 293 098   e 234 096 098

2 Round the nearest hundredth.
   a 2.995   b 7.399   c 4.961   d 1.0109   e 23.8476

3 Round 139 876 324 987 to the nearest billion.

4 Round each decimal to the nearest whole number and estimate the answer.
   a 138.22 + 137.66 + 138.32   b 44.501 + 80.89
   c 10.17 + 10.2 + 9.99 + 9.56 + 10.45

5 Estimate the answers to these calculations using the most efficient method.
   a 1 703 + 1 715 + 1 686   b 17 899 − 15 211
   c 54 999 ÷ 22   d 23 408 × 31

6 Estimate the answer and then solve each problem.
   a A hotel orders 5 400 litres of water in 45 litre containers. How many containers did the hotel order?
   b A builder orders 2 525 kg of cement in 12 kg bags. How many bags is this?
   c What is the area of a rectangular plot measuring 98 m by 385 m?

Topic 10 Mental Methods Workbook pages 25–27

▲ What types of numbers can you see in the photograph? Which of the sale offers means the same as half price? How would you work out the sale price of an item with 50% off? What about 25% off?

In our daily lives, we often have to do calculations in our heads; in other words, without using pen and paper methods and without using a calculator. In mathematics, knowing how to work out answers in your head is a valuable skill and it can help you work more quickly and efficiently to solve problems. In this topic, you are going to revise some of the **strategies** that you can use to add, subtract, multiply and divide **mentally**. You will also develop some mental methods for finding **percentages** of an amount.

Getting Started

1 What must be added to each number to make 10?

   a 3   b 9   c 8

   d 4   e 5   f $4\frac{1}{2}$

2 How can you quickly find half of these amounts?

   a 20   b 42   c 68

   d 110   e 250   f 101

3 How does knowing that 3 + 6 = 9 help you to calculate these sums?

   a 60 + 30   b 300 + 600   c 23 + 26   d 103 + 206

4 How can you tell whether a number is divisible by:

   a 2   b 5   c 10

   d 5 and 10   e 100   f 2, 5 and 10?

Unit 1 Mental Strategies

Let's Think …

- Marcus has to work out 5×28 in his head. This is how he thinks:
  Is his answer correct?
  What strategy did he use to calculate the answer?
- Michelle has to calculate 12×25. This is how she thinks:
  Is her answer correct?
  Why do you think she halved the 12 and doubled the 25?
- How would you work out 6×18 in your head? Write a thought bubble to show your thinking.

5 is half of 10.
10×28 is 280.
Half of 280 is 140.

12×25 is hard.
6×50 is easier.
6×5 is 30.
30×10 is 300.

A *mental* calculation is one that you work out in your head, by thinking, rather than using pen and paper methods. Sometimes you might need to jot down steps on scrap paper. This is not the same as doing long calculations; *jottings* are just reminders, not workings.

There are some *strategies* that you can use to help you do calculations mentally.

Read through these *strategies* carefully.

Type of Calculation	Strategy	Example
Multiply by 10 or 100	Put one or two zeros at the end of the number	$23 \times 10 = 230$ $23 \times 100 = 2\,300$
Multiply by 2	*Double* the number Add it to itself	$123 \times 2 = 123 + 123 = 264$
Multiply by 5	Multiply by 10 and *half* the answer	5×23 $10 \times 23 = 230$ half 2 30 half 100 15 115
Divide by 2	Half the values in each place or group of places	$518 \div 2$ 5 18 half half 250 9 259
Add two-digit numbers	Add the tens and then the units	$23 + 34 = 50 + 3 + 4 = 57$ $25 + 36 = 50 + 5 + 6 = 61$
Multiply by single digit numbers	Use doubling and halving where you can	5×46 is the same as $10 \times 23 = 230$ 7×18 is the same as $7 \times 9 \times 2 = 63 \times 2 = 126$

Type of Calculation	Strategy	Example
Subtract two-digit numbers	Subtract in parts if you can. Use chunking to count up and find the difference (think of a number line with jumps)	$73 - 21$ $73 - 20 \rightarrow 53 - 1 = 52$ $84 - 26$ $26 + 50$ is 76 $76 + 4$ is 80 and $80 + 4 = 84$ So the answer is $50 + 4 + 4 = 58$

You can also use mental methods to work out *percentages of amounts* if you remember that a percentage is a fraction of the whole.

50% is the same as $\frac{1}{2}$. To find a half, you divide by 2.

25% is the same as $\frac{1}{4}$. To find a quarter, you divide by 4, or divide by 2 and then 2 again.

10% is the same as $\frac{1}{10}$. To find a tenth, divide by 10.

20% is the same as $\frac{1}{5}$. To find a fifth, divide by 5 (or divide by ten and double the result).

1 Use the mental strategies you find easiest to do these calculations.

a $43 + 20$ b $64 + 20$ c $73 + 30$ d $29 + 60$

e $43 + 21$ f $63 + 42$ g $71 + 19$ h $34 + 43$

i $36 + 15$ j $47 + 26$ k $37 + 35$ l $43 + 58$

2 Calculate these mentally.

a $43 - 10$ b $65 - 30$ c $87 - 20$ d $49 - 30$

e $49 - 21$ f $67 - 23$ g $73 - 23$ h $98 - 43$

i $83 - 25$ j $63 - 35$ k $72 - 46$ l $54 - 38$

3 Calculate these mentally.

a 23×10 b 19×100 c 45×10 d 123×100

e 32×5 f 12×18 g 25×8 h 32×5

i 14×25 j 20×23 k 14×15 l 84×50

4 Calculate.

a 50% of $\$84.00$ b 10% of 120 c 10% of $\$450.00$ d 25% of 420

Looking Back

Noleen says to multiply by 25 mentally, you can times by 100 and then half the answer and half it again.

Test Noleen's method on a few examples.

Explain why it works.

Topic Review

What Did You Learn?

- Mental calculations are done in your head.
- There are different strategies that you can use to add, subtract, multiply and divide mentally.
- You learned a number of tips and strategies for calculating mentally.
- You can find a percentage of an amount by thinking of the percentage as a fraction and dividing.

Talking Mathematics

Discuss and then answer these questions in your mathematics journal.

- Did you learn any new methods of working in this topic?
- Which strategies do you find easiest to use? Why?
- Did you struggle with any of the strategies? Why?
- Why is mental mathematics useful?
- Besides school work, can you think of three examples where you might do a mental calculation?

Quick Check

1 Calculate mentally. Write the answers only.

a 50×38 b 17×50 c 40×16

d 56×25 e 76×25 f 25×32

2 What is:

a 47 less than 73   b 56 less than 93   c 25% of 84

d 43 more than 32   e 48 more than 19   f 10% of 920?

3 What is the value of x in each calculation?

a $x + 18 = 45$ b $x - 20 = 47$ c $23 + x = 100$ d $85 - x = 50$

4 Look at the prices of different tours.
Work out the price of these.

a Fishing fun and cave visit.
b Island hopper and cave visit.
c Chocolate and rum factory for 5 people.
d Cave visit for 20 people.
e A combined island hopper and fishing fun with 10% off the total.

Tours	per person
Fishing Fun	$280.00
Cave Visit	$170.00
Island Hopper	$450.00
Chocolate & Rum Factory	$180.00

Topic 11 Factors and Multiples Workbook pages 28–30

Workbook pages 28–30

Key Words

multiple

factor

product

common factor

common multiple

greatest common factor (GCF)

least common multiple (LCM)

prime number

prime factors

▲ Count the beads in each pack. What is the least number of packs you would need to buy if you wanted to get the same number of green beads and orange beads?

Last year, you learned about **factors** and **multiples** and you listed the factors and multiples of different numbers to compare them and find the **greatest common factor (GCF)** and **least common multiple (LCM)**. You also learned about prime numbers and you used factor trees to break numbers down so that you could write them as the **product** of prime factors. In this topic, you are going to revise some of these concepts and learn how to use prime factorization to find the GCF and LCM of two or more numbers.

Getting Started

1 Work in pairs to write a definition and give an example of each of the following mathematical terms.

   a factor

   b multiple

   c prime number

   d prime factor

   e product of prime factors

2 What number is a factor of every whole number above zero?

3 How would you describe each of these sets of numbers? Why?

   a 1, 2, 3, 4, 6, 8, 12, 24

   b 3, 6, 9, 12, 15, 18

4 If a person's age is a multiple of two and also a square number less than 25, how old could they be?

Unit 1 Revisit Factors and Multiples

Let's Think ...

One way of writing 75 as a product is 1×75.
How many other ways can you find?
Explain why you cannot divide 75¢ equally among seven people.

The *factors* of a number are all the whole numbers that divide into it exactly.
The factors of 12 are 1, 2, 3, 4, 6 and 12.

Some numbers share factors. Shared factors are called *common factors*.
The purple factors are common to 12 and 16.

Factors of 12:   1, 2, 3, 4, 6, 12
Factors of 16:   1, 2, 4, 8, 16

The *greatest common factor* or *GCF* of 12 and 16 is 4.

Multiples of a number are found when you multiply that number by any other whole number.
The first ten multiples of 3 are: 3, 6, 9, 12, 15, 18, 21, 24, 27, 30
The first ten multiples of 4 are: 4, 8, 12, 16, 20, 24, 28, 32, 36, 40

12 and 24 are multiples of both 3 and of 4. We call these *common multiples*.
The lowest number that is common to both sets is 12.
We say that 12 is the *least common multiple* or *LCM* of 3 and 4.

Prime numbers are numbers with two (and only two) factors. The factors are 1 and the number itself.
7 is a prime number. The only two whole numbers that can divide exactly into 7 are 1 and 7 itself.
The factors of 7 are 1 and 7.

All numbers can be written as a *product of prime numbers*.

To find the *prime factors* of a number you can use a prime factor tree.

Write 60 as a product of prime factors.

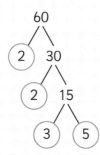

Write the number as a product of any two factors.
Circle any primes.
Write any non-prime factors as a product of two numbers.
Continue until all factors are prime.
Write the product using multiplication signs or powers.

$2 \times 2 \times 3 \times 5$
$2^2 \times 3 \times 5$

1 List the first ten multiples of:
   a 4   b 8   c 7   d 9   e 11

2 List three numbers that are common multiples of:
   a 2 and 3   b 5 and 6   c 4 and 5   d 9 and 12

3 What is the LCM of each set of numbers?
   a 4 and 6   b 3, 4 and 8   c 12 and 8   d 2, 5 and 10

4 Find all the factors of:
   a 32   b 16   c 80   d 100   e 36

5 By listing the factors, find the HCF of each set of numbers.
   a 12 and 16   b 18 and 40   c 40 and 60   d 38 and 39

6 Say whether each statement is TRUE or FALSE.
   a 4 is a factor of 25.
   b 6 is a factor of 84.
   c 7 is a factor of 56 and 84.
   d All multiples of 6 are also multiples of 3.
   e All multiples of 4 are also multiples of 8.
   f The product of two prime numbers is always even.
   g The sum of two prime numbers is always even.

7 Express the following as a product of prime factors.
   a 14   b 32   c 40   d 36   e 100
   f 156   g 225   h 80   i 24   j 1 000

8 Sally, James and Keshorn are checking their email accounts to see if an invitation to a party has
   arrived. Sally checks hers every 4 minutes, James checks his every 6 minutes and Keshorn checks his
   every 14 minutes.
   a After how many minutes will Sally and James check their accounts at the same time?
   b After how many minutes will all three of them check their accounts at the same time?

9 Toniqua has 60 red beads and 90 white beads. She wants to use all the beads to make identical
   bangles. Each bangle should have the same number of red and the same number of white beads. What
   is the greatest number of bangles she can make?

10 Nicky has two lengths of ribbon. One is 32 cm long and the other is 60 cm long. She wants to cut as
    many pieces of equal length as possible from the two ribbons. How long will each piece be?

Looking Back

Marcus has written some incorrect factors in each list. Which are they?

Factors of 24:   4   2   6   8   1   16   24   3

Factors of 23:   3   1   7   23

How can working in order with factor pairs help you to avoid mistakes like Marcus made?

Unit 2 Use Prime Factors to Find the GCF and LCM

Let's Think ...

24 can be written as a product of its prime factors like this:

   $2 \times 2 \times 2 \times 3 = 24$

Work in groups.

Investigate how you could use these prime factors to work out all the factors of 24 (there are 8 in total).

It can take a long time to list multiples or work out factor pairs in order to find the least common multiple or highest common factor.

Writing a number as a product of its prime factors saves time and allows you to find the LCM and GCF efficiently.

You can either use the factor tree method you already know, or you can use division, to find the product of prime factors.

Example 1

Express 84 as a product of its prime factors.

Lowest prime → $\begin{array}{r|r} 2 & 84 \\ \hline 2 & 42 \\ \hline 3 & 21 \\ \hline 7 & 7 \\ \hline & 1 \end{array}$ ← Next lowest prime

Divide the given number by the lowest prime number that divides exactly into it. Always try 2 first.
Divide the quotient by its lowest prime factor.
Move to the next prime number if necessary.
Continue until you get a quotient of 1.
Write the divisors as a product.

   $84 = 2 \times 2 \times 3 \times 7$

Example 2

Find the GCF of 24 and 60.

Use factor trees or division to find the prime factors.

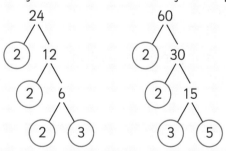

$24 = ② \times ② \times 2 \times ③$ Mark the common factors.
$60 = ② \times ② \times ③ \times 5$ $2 \times 2 \times 3$ is common to both sets.

Find the product of the common prime factors: $2 \times 2 \times 3 = 12$

The GCF is 12.

Example 3

Find the LCM of 8 and 14.

Use factor trees or division to find the prime factors.

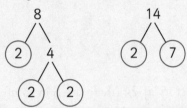

$8 = \textcircled{2} \times \textcircled{2} \times \textcircled{2}$ Find where each factor appears the most times and mark the factor each time it appears in that set.

$14 = 2 \times \textcircled{7}$ The first set has most 2s, the second set has most 7s.

Find the product of the marked factors: $2 \times 2 \times 2 \times 7 = 56$

The LCM is 56.

---

1   Use prime factors to find the GCF of each pair of numbers.

> *If you have worked out the product of prime factors in one part of the exercise, you can reuse it if the number appears in another set. You do not need to work it out again.*

   a   15 and 20   b   12 and 16   c   20 and 16   d   30 and 36

   e   25 and 35   f   10 and 20   g   18 and 30   h   36 and 63

2   Write each number as a product of its prime factors and write the GCF of each set.

   a   75, 120 and 150   b   24, 40 and 80   c   12, 48 and 60

3   Use prime factorization to find the LCM of each pair of numbers.

   a   15 and 18   b   24 and 28   c   22 and 25   d   18 and 24

   e   25 and 30   f   72 and 108   g   95 and 120   h   22 and 33

4   Determine both the GCF and the LCM of each pair of numbers using prime factorization.

   a   36 and 60   b   36 and 48   c   60 and 80   d   52 and 78

Looking Back

1   a   Find the LCM of 2, 3, 4 and 6.

    b   What is the GCF of 2, 3, 4 and 6? Why?

2   Use prime factors to find the LCM and GCF of 8, 12, 16 and 30.

Topic Review

Talking Mathematics

These statements are all incorrect. Explain why they are incorrect.

- The LCM of two numbers is the lowest number that divides into them both with no remainder.
- To find the GCF of two numbers you just multiply them.
- The LCM of 4 and 6 is 24 because $4 \times 6 = 24$.
- The highest common factor of two prime number is the sum of the numbers.

Quick Check

1 Say whether each statement is TRUE or FALSE.
   a Common multiples of 3 and 5 must be odd numbers.
   b All common multiples of 2 and 5 are also multiples of 10.
   c The GCF of 5 and 11 is 1.
   d 34 has four ordinary factors and two prime factors.
   e The smallest number you can find with exactly three factors is 9.

2 List:
   a all the factors of 36
   b the prime factors of 36
   c the four lowest multiples of 36
   d the factors of 36 that are square numbers.

3 Find the difference between the sum of the prime numbers less than 10 and the sum of the prime numbers between 10 and 20.

4 Use prime factors to find the GCF and LCM of 12, 15 and 18.

Topic 12 Scale and Distance Workbook pages 31–33

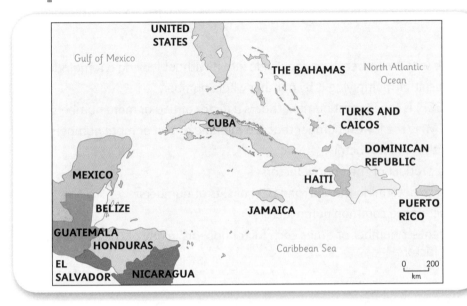

▲ What do we call this kind of diagram? What does it show us? How can you work out the distance between two places on this diagram?

Maps are graphic representations of the real world. They show us where places are in relation to each other. Maps are much smaller than the real areas that they show, so we need a method for working out how far two points on the map are from each other in the real world. In this topic, you are going to learn how to read and make sense of different types of **scales** on maps. You will use the scales to calculate the real distance between places shown on a map.

Getting Started

1 Look at these pictures of three famous structures from around the world.

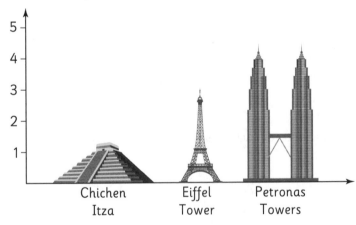

a Why are diagrams like this called scaled diagrams?

b If 1 division on the diagram scale represents 100 m in the real world, estimate the real height of each structure.

c How many times bigger are the heights in the real world?

Unit 1 Scales on Maps

Let's Think ...

Measure the lengths of the sides of your desk to the nearest centimetre.
Use a scale of 1 cm to show 10 cm of desk to draw a scale drawing of your desk.

When you draw a map, you have to draw things smaller than they really are to fit onto the page. The distances on the map are a *fraction* of the real distances. The map *scale* tells you how much smaller the map is than the real world.

$$\text{The scale of the map} = \frac{\text{length on the map}}{\text{length in the real world}}$$

Your scaled drawing of your desk used a scale of 1 cm on the map for 10 cm on the desk.

$$Scale = \frac{1\,cm}{10\,cm}$$

The scale drawing is therefore $\frac{1}{10}$ of the size of the real desk.

The scale of a map can be given in words, for example 1 cm = 5 km. This means that 1 cm on the map represents 5 km in the real world.

Scale can also be expressed as a *ratio* such as 1 : 1 000, which means that 1 unit on the map represents 1 000 of the same units in the real world. So, 1 cm on the map represents 1 000 cm in the real world and 1 mm on the map represents 1 000 mm in the real world.

On many maps, you will find a *scale bar*.

```
0        100       200       300 Kilometres
├─────────┼─────────┼─────────┤
0              100       200       300 Miles
```

The small divisions on the scale bar show map distances, but the number tells you what the real distances are. You can compare distances on the map with the scale bar to work out how far they are in reality.

Look at the map on page 68. The bar scale shows that 1 cm on the map represents a real distance of 50 km.

What is the approximate distance between Freeport and Nassau?

1   Express the following scales in words:

a

```
0        5        10       15       20 km
```
b   1 : 25 000

c
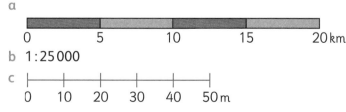
```
├──┼──┼──┼──┼──┤
0   10  20  30  40  50 m
```

2   Refer to the map of The Bahamas on page 68. Pick any three pairs of locations, measure the distance between them in centimetres and then calculate the actual distance between them. Write down your results.

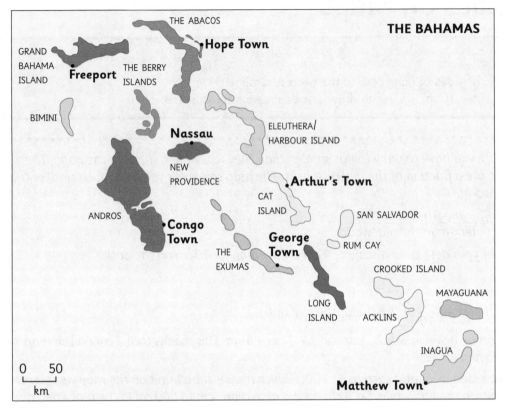

Using a ruler to measure the direct distance between Freeport and Nassau on the map, you can see that it is just about 4 cm.

Therefore, the actual distance between the two cities is more or less

4 × 50 km = 200 km.

3 On the right is a map that shows the relationship between three places in a town. Answer the questions that follow, based on the map.

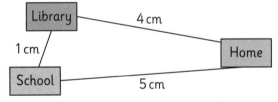

Scale: 1 cm = 1 km

a Michael walks to the library after school each day. How far does he walk to get there?

b He then walks back to school to meet his sister and they walk home together. How far does he walk this time?

c On Tuesday, Michael's sister was not at school so he went straight home from the library. What was his total distance on Tuesday?

Looking Back

- Place some objects on your desk. Use the drawing you made of your desk before and draw in these objects to scale.
- Add a scale to your drawing.
- Check your scale drawings with a partner.

Topic Review

Talking Mathematics

- Explain what a scale is and how it is used on maps.
- What problems could people have if they use maps without a scale? Why?

Quick Check

1 The scale of a map is 1 : 20 000. What distance on the map will represent 2 km?

2 A map uses a scale of 1 cm = 20 miles. Two towns are 6.5 cm from each other on the map. What is the actual distance between the two towns?

3 You want to add a location to a map with a scale of 1 cm = 15 km. The location is 45 km from the city. How many centimetres away from the city will you mark the location?

4 Use the map to find the distance between the following places.

   a Airport and town centre
   b Airport and mountain cabin
   c Restaurant and picnic spot
   d Harbour and restaurant
   e Mountain cabin and harbour

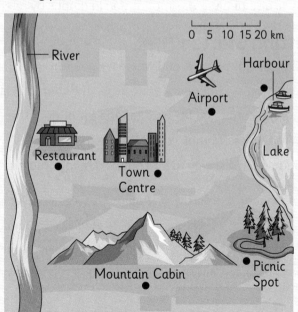

Topic 13 Graphs Workbook pages 34–38

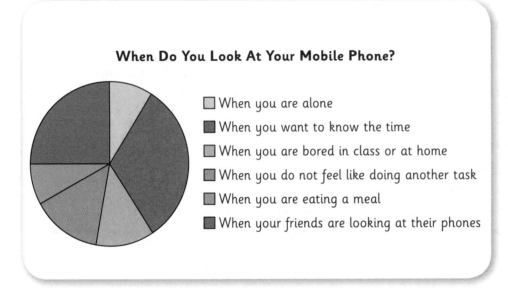

When Do You Look At Your Mobile Phone?

☐ When you are alone
◼ When you want to know the time
◼ When you are bored in class or at home
◼ When you do not feel like doing another task
◼ When you are eating a meal
◼ When your friends are looking at their phones

▲   This circle graph shows when a group of students look at their phones. Does this relate to the way you use your own phone? How is your usage different or the same?

Graphs are used to display information or data visually. Before you can draw a **graph**, you have to collect relevant **data** and decide which type of graph would be best to represent the information.

In this topic, you are going to revise ways of collecting data and how to represent it using different graphs. You will also analyse the information in graphs by referring to the **mean**, **median**, **mode** and **range**.

Getting Started

1   Explain what data is and where we get it from.

2   Why would a frequency table be useful when collecting data?

3   Are the mean and the median of a set of data the same thing?

4   What types of graph could you use to compare two sets of data?

5   Shania did a survey to find out how many students had a birthday in each month. These are her results:

January	March	August	November	December	June	August	June
June	February	June	August	January	November	September	August	February
July	April	May	November	September	August	January	October
November	October	July	April	April	May	December

a   How many people did she collect data from?

b   Why is this data not so easy to work with?

c   Draw a table to summarize this data and make it easier to work with.

Unit 1 Organizing and Representing Data

Different types of graphs can be used to display different types of data.

Pictographs use pictures or symbols to show and compare data that can be counted. They show how many or how much. A pictograph must have a key to show what each picture or symbol represents.

Bar graphs use bars to show and compare countable data. The graph can be vertical or horizontal. A dual or double bar graph shows and compares two similar sets of data.

Circle graphs (or pie charts) show how data is divided or shared. It shows how one part of the data is related to another part and also how it is related to the whole set of data.

Line graphs show how data changes over time. This sort of data is called continuous data. A line graph can show a trend or direction.

1   Which types of graphs could you use to display the following sets of data? Discuss your answers.

   a Data that shows how many days of sunshine there are in The Bahamas each month of the year.

   b Data that shows the time that people in different countries spend on social media each day.

   c Data that shows the body temperature of a patient in a hospital every hour.

   d Data that shows how much money a family spends on different items each month, as a percentage of what the family spends altogether.

2   Study the pictograph quickly. It shows the number of swimmers who were swimming each day at a beach at 12:00. Answer the questions, without counting.

Monday 12:00	🏊 🏊 🏊 🏊 🏊 🏊
Tuesday 12:00	🏊 🏊 🏊 🏊 🏊 🏊 🏊 🏊 🏊
Wednesday 12:00	🏊 🏊 🏊 🏊
Thursday 12:00	🏊 🏊
Friday 12:00	🏊 🏊 🏊
Saturday 12:00	🏊 🏊 🏊 🏊 🏊 🏊 🏊 🏊 🏊 🏊 🏊 🏊 🏊 🏊 🏊
Sunday 12:00	🏊 🏊 🏊 🏊 🏊 🏊 🏊 🏊

Key: 🏊 = 3 swimmers

a Which day had the most swimmers at 12:00?

b On which days were there just a few swimmers?

c On which days were there the same numbers of swimmers?

d Why are pictographs useful?

3 This graph shows the number of followers that some well-known sport stars have on a popular social networking service. The numbers are in millions.

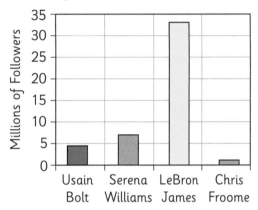

a Which of these sport stars has the most followers?

b Approximately how many followers does Chris Froome have?

c What conclusions could you draw from this data? Discuss your ideas with a partner.

4 The circle graph shows how students in a class get to school each day. The graph has four *sectors*.

Transport

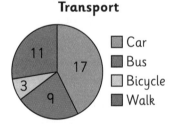

☐ Car
☐ Bus
☐ Bicycle
☐ Walk

a Name each sector.

b What is total number of students represented in this data?

c How can fractions help you estimate what each sector of a circle graph represents?

5 Study the double bar graph and answer the questions.

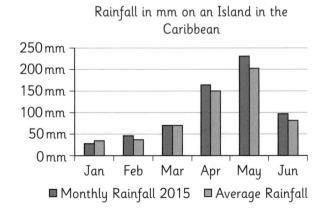

Rainfall in mm on an Island in the Caribbean

☐ Monthly Rainfall 2015   ☐ Average Rainfall

a What statistical data does this graph show?

b In which months was the monthly rainfall higher than average?

c In which months was the monthly rainfall lower than average?

d For what other types of data would a graph like this be useful?

6 A clothing shop used sales data to compile this graph of sales for the first 6 months of the year. The sales figures are in US $, in thousands. Study the graph and discuss the questions.

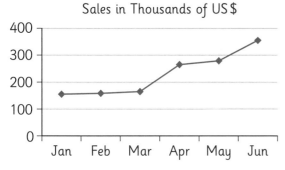

Sales in Thousands of US$

a Should the management of this shop be pleased or unhappy?

b Why is a graph like this useful to shop management?

c Would it be useful to compare sales with sales at other times? How could that be done? What could be compared?

7 Look at the following sets of data. Then discuss with a partner the type of graph you could use to show this data. You can draw these graphs in your Workbook.

a Tori wanted to find out which television programmes people in her grade watched. She interviewed 30 people and recorded the data in this frequency chart. Use the data to draw a graph.

TV Programme	Frequency
Local news channels	5
International news channels	3
Sports channels	12
Series	6
Movies	4

b The following table shows the temperature of the water in the sea at different times during one day. Draw a graph to represent this data. Use the graph to estimate the temperature of the sea at 12:30 and at 14:30. Comment on the temperature changes during the day.

Time	9:00	10:00	11:00	12:00	13:00	14:00	15:00
Temperature (in °C)	21	22	24	26	28	29	29

Looking Back

Work in pairs. Look through some of your local newspapers. Find examples of different graphs. Describe to your partner the data that is shown in each graph that you find.

Unit 2 Collecting Data

Let's Think ...

- Look at the circle graph. What does it tell you?
- How could the person who drew the graph have collected the information?
- How do we know that data used in surveys is accurate?

Things People Do While Making Toast

- ☐ Get a plate ready
- ☐ Clean up the kitchen
- ☐ Stare at the toaster

You can collect data or information from many sources; for example, from surveys, from sales figures or from making observations.

If you need to collect data, you can conduct a survey. You will need to create a questionnaire and record the results. You can do this using a frequency table (also called a tally chart). You can then draw a graph and analyse the data you have collected.

1 Choose the better question for each survey. Say why the other question is not suitable.

   a A survey to find out what food is the most popular in the school canteen.

- Do you buy food from the school canteen?
- What food do you buy at the canteen?

   b A survey to find out about favourite colours.

- What colour do you like?
- Do you like pink or blue best?

2 Data was collected for what food students bought in the school cafeteria one day. Draw a frequency table (tally chart) to show the data.

sandwich	cookies	cold drink	chocolate	cookies
fruit	fruit	sandwich	cold drink	cookies
cookies	cold drink	fruit	chocolate	sandwich
sandwich	cookies	cold drink	fruit	cookies
sandwich	cookies			

3 Your school wants to encourage students to take part in sport. They want to introduce new sports at the school. Work in groups. Decide how you could help the school to make a decision about which sports to introduce. You will need to collect data and make a presentation to back up your suggestions.

Looking Back

Copy and complete the following flow diagram to show how to collect and organize data.

Create questions.	⇨		⇨		⇨	

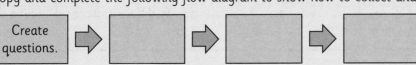

Unit 3 Analyse Data (Mean, Mode, Median and Range)

Let's Think …

What does the word 'average' mean in these statements?

• The average score for the test was 22.

• The average temperature in December was 25 °C.

• Young children in this town spend an average of 3.5 hours a day watching television.

An average is a value that is typical of a set of data. There are different types of averages in statistics: the *mean*, the *mode* and the *median*.

The *range* of a set of data is the difference between the highest and the lowest values.

Value	How to Calculate	Example
Mean	Add all the values of the numbers in a set and then divide the total by the number of values in the set.	Set: 4, 6, 8, 5, 3, 9 $4 + 7 + 8 + 5 + 3 + 9 = 36$ $36 \div 6 = 6$ **The mean is 6.**
Mode	Find the number that occurs the most often.	Set: 7, 7, 8, 2, 11, 10 7 occurs the most often **The mode is 7.**
Median	Put the numbers in order and find the middle number.	Set: 6, 4, 10, 6, 9, 3, 8 In order: 3, 4, 6, 6, 8, 9, 10 **The median is 6.**
	If there are two middle numbers that are not the same, find the mean of the two numbers.	Set: 14, 20, 17, 18, 21, 15 In order: 14, 15, 17, 19, 20, 21 Middle numbers: 17, 19 $17 + 19 = 36$ $36 \div 2 = 18$ **The median is 18.**
Range	Subtract the lowest number from the highest number in the set.	Set: 4 6 8 5 3 9 $9 - 3 = 6$ **The range is 6.**

1 Find the mean of each set of data.

a 11   16   12   19   9

b 15 mm   21 mm   18 mm   19 mm   23 mm

c $2.50 $3.50 $4.00 $7.00 $1.00

d 36 °C   37 °C   30 °C   32 °C   34 °C   33 °C

2 Study the data in the table.
   a What is the mean?
   c What is the median?
   b What is the mode?
   d What is the range?

Student	Mathematics Test: Mark out of 20
Terri	10
Bill	11
Sierra	18
Mario	15
Jesse	14
Ashley	19
Torianne	12
Ernest	13
Micah	14
Rachel	14

3 Study the set of data about the prices of cricket bats.

| Prices: $41.00 $79.00 $34.00 $70.00 $34.00 $68.00 $59.00 |

Choose the correct answers.

a The mean is …
   $54.00 $45.00 $55.00 $70.00
b The mode is …
   $54.00   There is no mode. $29.00 $34.00
c The median is …
   $34.00 $59.00   There is no median. $55.00
d The range is …
   $45.00 $59.00 $34.00 $70.00

4 A football club needs to choose the best strikers for a competition.
   They decide to look at the goals scored by each striker in the last
   5 games. Here is the set of data.

   a Which striker had the highest average number of goals? What is
      the mathematical term for this? What mathematical calculation
      did you do?

   b Based on this data, who is the worst striker? Why?

   c Which strikers should they choose? Why?

Name	Goals
Kirkland	0 1 0 2 2
Jo	1 1 2 1 0
Ryan	2 1 2 2 3
Ahkeem	0 1 2 0 0
Benoni	1 0 0 1 1

Looking Back

Copy and complete the sentences.
a The ___ of a set of data is the sum of the values, divided by the number of values in the set.
b The mode of a set of data is ___.
c To find the ___ of a set of values, you first have to arrange the values in order.
d The range of a set of data is the ___ between the highest and the lowest values.

Topic Review

Talking Mathematics

What is the mathematical word for each of these?

- A graph that shows information for two related sets of data.
- A graph that uses pictures to represent data.
- A graph that uses bars to show the values of data.
- A graph that shows how values are shared or divided.
- The value in a set of data that occurs most often.
- The value that tells you how widely a data set is spread.

Quick Check

1 Which type of graph shows you trends or directions?

2 When would you use a double bar graph to show data?

3 What does a circle graph show?

4 Draw a rough circle graph to show how much time you spend sleeping, eating, attending school and doing other things each day. Label your graph clearly so other people can understand it.

5 Name three different types of averages that you can calculate in mathematics.

Topic 14 Adding and Subtracting Workbook pages 39–40

10th Annual Coast-to-Coast Fun Run

Start

Stage 1: 2805 m

Stage 2: 2290 m

Stage 3: 995 m (UPHILL!)

Stage 4

Finish

An 8000 m trail run

Key Words

estimate

compatible

add

sum

subtract

difference

regroup

rename

▲ Read the information about the Fun Run. How long is the whole run? How far will you have run if you have completed the first three stages? The length of Stage 4 is not given. How could you work it out?

You already know how to **estimate** using rounding and clustering and you can **add** and **subtract** large numbers. You should also remember the mathematical language that tells you whether to add or subtract. In this topic, you are going to learn how to estimate using **compatible** numbers and then you are going to practice your skills and apply them to solve problems in which you add and/or subtract to find the solution.

Getting Started

1 Use the figures above to work out the following.
   a The difference in distance between Stage 1 and Stage 2.
   b The total distance from the end of Stage 1 to the start of Stage 4.
   c How much further you run in Stage 2 than in Stage 3.

2 Joe trains for the race by running the Stage 3 route four times each day for a week. Micah trains by running the Stage 1 route twice a day for a week.
   a Who runs further?
   b How much further do they run?

3 A school decides to use the fun run to raise money for charity. Students get sponsors who pay for each kilometre they run. Micah gets three people to sponsor him. One will pay $5.00 per kilometre, one will pay $8.00 per kilometre and the other will pay $9.00 per kilometre. If Micah finishes the entire race, how much money will he raise?

Unit 1 Revisit Addition and Subtraction

Let's Think ...

Haiti and Dominican Republic are two countries on
the same island. The area of Haiti is 27 560 square kilometres
and the area of Dominican Republic is 48 320 square kilometres.

a What is the total area of the island?

b What is the difference between the areas of Haiti and
Dominican Republic?

c Tell your partner how you worked out your answer.

You already know how to *add* and *subtract* numbers using expanded notation and column methods.

You also know that it is important to *estimate* before you add or subtract so that you can decide
whether your answer is reasonable or not. You can round numbers to a suitable place value to estimate
and you can estimate the *sum* of several similar values by clustering.

You can also estimate sums and *differences* using compatible numbers.

A compatible number is one that is fairly close to the value you are working with, but is easier to add or
subtract.

Compatible numbers can give you an estimate that is closer to the real value than rounded numbers.

Example 1

a $455 + 223 \approx 670$

$450 + 220 = 670$

250 and 220 are compatible so you can add
them mentally.

Rounding the values to the leading place would
give $500 + 200 = 700$. The compatible numbers
give 670, which is closer to the real value of 678.

b $845 - 637 \approx 200$

$850 - 650 = 200$

850 and 650 are compatible so you can
subtract them mentally.

Read through the next two examples if you have forgotten how to add and subtract larger numbers in columns.

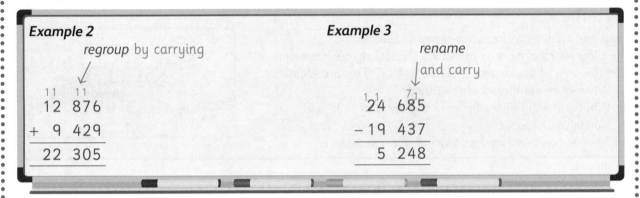

Example 2

regroup by carrying

$$\begin{array}{r} {}^{11}12\ {}^{11}876 \\ +\ 9\ 429 \\ \hline 22\ 305 \end{array}$$

Example 3

rename and carry

$$\begin{array}{r} {}^{1}2{}^{1}4\ {}^{7}6{}^{1}85 \\ -\ 19\ 437 \\ \hline 5\ 248 \end{array}$$

1 Estimate and then add.

   a $142 + 117 + 131$ b $289 + 2114 + 1309$ c $12345 + 4568 + 99$

   d $1456 + 7654 + 8123$ e $765 + 1234 + 678 + 99$ f $43568 + 12897$

2 Estimate and then subtract.

   a $8798 - 1307$ b $8032 - 4156$ c $12345 - 8765$

   d $321098 - 158987$ e $88950 - 43999$ f $121625 - 112887$

3 The table shows the area and population (2016) of each continent.

Continent	Area in Square Kilometres	Population (2016)
Asia	43 820 000	4 436 224 000
Africa	30 370 000	1 216 130 000
North America	24 490 000	579 024 000
South America	17 840 000	422 535 000
Europe	10 180 000	738 849 000
Australia	7 690 500	39 901 000
Antarctica	13 720 000	4 490

   a How much bigger is Asia than Europe?

   b How much smaller is South America than North America?

   c What is the difference between the greatest and least populations?

   d What is the combined area of the Americas?

   e What is the combined population of the Americas?

Looking Back

Make up two addition problems and two subtraction problems using the data from the table.
Exchange with another student. Check each other's answers when you have solved the problems.

Unit 2 Mixed Problems

You can use addition and subtraction to solve many different kinds of mathematical problems. However, before you do any calculations, you need to read the problem and decide which operation or operations you need to do.

Sometimes you need to do more than one operation to solve a problem.

Macie buys old furniture and sells it to make a profit. The difference between what she pays and what she sells it for is her profit.

Macie paid $420.00 for two desks. She sold one for $348.00 and the other for $364.00. What was her profit on the two desks?

Think: Her profit is what she sold them for minus what she paid.

Sold for:	Minus what she paid:
$348.00	$712.00
+364.00	−420.00
$712.00	$292.00

Her profit was $292.00

Before you tackle these problems, read them carefully and work out what operation you need and whether you need to do more than one calculation.

1   Macie also bought a table for $329.00 and a chair for $89.00. She later sold them as a set for $700.00. What was her profit on that sale?

2   The marks out of 100 for four students in three exams are given below.

Student	English	Mathematics	Science
Jonelle	76	82	72
Keshawn	75	81	79
Kezia	81	79	77
Leroy	74	78	76

a   What is the total score out of 300 for each student?

b   What is the range of the total scores?

c   What is the mean score for each subject?

3 The length of some of the world's longest rivers is given below.

River	Amazon	Nile	Yangtze	Mississippi	Yellow
Length (km)	6992	6853	6418	6275	5464

a What is the range of lengths?

b The Yangtze and Yellow Rivers are both in China. What is their combined length?

c How far would you travel if you sailed all the way up and down the Nile?

d A researcher wants to travel the length of all five rivers. What is this distance?

4 The land area of some large islands is given here in square kilometres.

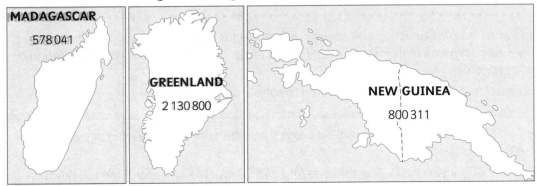

MADAGASCAR
578041

GREENLAND
2130800

NEW GUINEA
800311

UK
216777

BAFFIN
507451

CUBA
110860

a The area of The Bahamas is 10070 square kilometres. Work out how much bigger than The Bahamas each other island is.

b Which is larger: Baffin Island or Madagascar, and by how much?

c Which two islands have a combined area of 1 million square kilometres when it is rounded to the nearest million?

Looking Back

The solutions to three different problems are given below. For each one, write a word problem that fits the number sentences.

a 23456 − 12987 = 10469

b 463 + 587 = 1050
  1050 − 209 = 841

c 345 + 245 + 650 = 1240
  1845 − 1240 = 605

Topic Review

Talking Mathematics

Do this calculation: Subtract 2 134 from 5 000.

- Explain how and why you rename digits in calculations like this.
- How could you check your answer by addition?
- Why should you estimate before you add or subtract large numbers?
- What method of estimation do you think is best? Why?

Quick Check

1 Write each of these numbers in numerals.
   a two million seventy-three thousand four hundred thirty-seven
   b seven hundred twenty-one thousand eight hundred twenty-nine
   c twelve million eight hundred forty-seven
   d seven million three hundred twenty-three thousand four hundred two

2 Use the numbers in question 1.
   a What is the sum of a and c?
   b What is the total of all four numbers?
   c Find the difference between the sum of a and b and the sum of c and d.

3 Calculate.
   a 190 991 + 12 234 + 14 568   b 23 132 819 + 9 343 214
   c 312 112 345 + 876 123 145   d 113 285 − 29 873
   e 3 229 876 − 1 314 388   f 8 234 000 − 1 939 453

4 Round to the nearest hundred thousand and estimate the sum of 911 270, 522 701, 654 688 and 813 431. Calculate the difference between your estimate and the actual total.

5 The sum of three numbers is 12 142 345. Write four different addition sums that will give this result.

6 The difference between two nine-digit numbers is 4 312 876. What could the numbers be?

▲ Mr Johnson grew one of these vegetables. The one he grew was not red or green and it was not round. Which one did he grow? What strategy did you use to work out the answer?

To solve problems, you first need to read and understand the problem, then you have to choose a suitable **strategy** to **solve** it. Next, you work out the solution and, finally, you check that it seems reasonable and correct. Last year, you learned that you can use **equations** to solve problems. In this topic, you are going to solve **non-routine** problems using the most suitable strategies. You are then going to revise **expressions** and equations and use them to solve problems.

Getting Started

1 Read the two problems.

   a Discuss what you need to find out in each one.

   b What steps do you need to follow to get a solution?

   c Try out your ideas. Can you solve the problems?

> 4 202 people attended a concert at the weekend. This is 11 times more than the number of musicians who performed at the concert. How many musicians were there?

> How many heart (♥) and diamond (♦) symbols are there in a complete pack of playing cards?

Key Words

strategy

non-routine

operations

create

solve

equation

expression

unknown

variable

inverse

Unit 1 Solve Non-Routine Problems

Let's Think …

Look at this plastic puzzle. The equilateral triangle fits into the space in the frame.
How many ways are there to put the equilateral triangle into the empty triangle shape?

 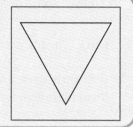

There is often more than one way to *solve* a problem. Many *non-routine* problems can be solved just by thinking and working systematically. Read through the *strategies* and the examples to see how to use each strategy to solve problems.

Look for Patterns

What happens to the area of a square if you double the length of the sides?

Side Length	1	2	4	8	16
Area (Side × Side)	1	4	16	64	256

Pattern: × 4   × 4   × 4   × 4

The area gets 4 times bigger if you double the length of the sides.

Trial and Error (or 'Guess, Check and Refine')

Dennis's age is $\frac{1}{5}$ of his dad's age. The sum of their ages is 54. How old is Dennis's dad?

Try 40   (Guess)

$\frac{1}{5}$ of 40 = 8   (Check)

40 + 8 = 48

That is too low, so try again   (Refine)

Try 45   (Guess)

$\frac{1}{5}$ of 45 = 9   (Check)

45 + 9 = 54

So Dennis's dad is 54.

Make a List, Chart, Table or Tally

How many whole numbers less than 100 contain the digit 5?

Five can be in the ones place: 5, 15, 25, 35, 45, 55, 65, 75, 85, 95

Five can be in the tens place: 50, 51, 52, 53, 54, X, 56, 57, 58, 59

Do not count 55 as it is in the first list.

10 + 9 = 19

Sketch a Drawing, Diagram or Model

I can see some goats and chickens in a field. There are 7 heads and 24 legs. How many goats are there?

Draw 7 heads

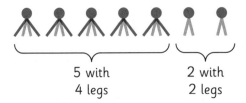

Give each head 2 legs
(they must all have at least 2)

⟶ 14 legs 'used'
10 legs 'left'

Share the 10 legs using
2 at a time

5 with
4 legs

2 with
2 legs

So there are 5 goats.

Eliminate Possibilities

Mike, Pete, James and Charles live in four-storey apartment block. The floors are numbered from 1st to 4th. Their surnames are Brown, White, Grey and Johnson, but not necessarily in that order.

Mike lives two floors below James.

Pete lives on the top floor.

Mr Brown lives on the third floor.

Mr White lives on the floor above Mr Grey.

What is the name and surname of the person on the second floor?

Jotting:

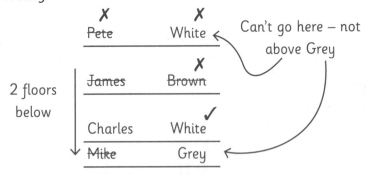

Pete and Brown can be eliminated immediately.

Mike must live on the first floor if he is two floors below James and James must live on the third floor. That leaves only Charles as a first name.

If Mr White lives above Mr Grey, he cannot live on the fourth floor, so he must live on the second floor.

The person on the second floor has to be Charles White.

Work in pairs or small groups. Decide which strategy is most useful for solving each problem and then work on your own to solve it.

1 Angela bought a present and a card for her friend. The present cost $16.00 more than the card and she paid $25.00 in total. What was the cost of the card?

2 This diagram contains 5 squares.

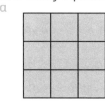

4 → 1 × 1 squares

1 → 2 × 2 square

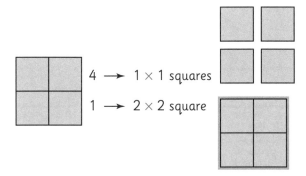

How many squares in each of these diagrams?

a

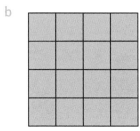

b

3 If everyone arriving at a party shakes hands with every other person there, how many handshakes will there be altogether if there are:

a 2 people   b 5 people   c 10 people?

d Can you find a method of working out how many handshakes there would be for *n* people?

4 Grandma Smith tells her grandson that she is 60 years old and that she married Grandpa Smith 34 years ago. She got a Master's Degree 3 years after she married. How old was she when she got her Master's Degree.

5 T-shirts on a stall are $8.60, $9.20, $10.80 or $12.40 depending on the size. A customer bought six T-shirts. Which of these amounts could be the total cost?

A $63.30   B $74.50   C $66.60

Looking Back

How many ways can you make $1.00 using any combination of 10, 20, 25 and 50 cent coins?

Unit 2 Expressions and Equations

Let's Think …

Write an equation to represent each problem and then solve it to work out the unknown value.

a 15 more than x is 20
b 10 less than a is 30
c 15 minus b is 10
d half of x is 22
e twice y is 30
f the square of x is 16

A mathematical expression is a group of numbers and/or letters linked by operations symbols.

These are all expressions:

$$3 + 5 \qquad x + 5 \qquad 6 - x \qquad y \times 4 \qquad \frac{10}{x}$$

The letters used in expressions are called variables. The value of an expression such as $x + 3$ can vary, depending on what number you use in place of x.

When $x = 1$, the expression has a value of $1 + 3$. When $x = 10$, the expression has a value of $10 + 3$.

An equation is a mathematical statement that makes two quantities equal. In other words, it is a number sentence with an equals sign.

These are all equations.

$$3 + 5 = 8 \qquad x + 5 = 9 \qquad 6 - x = 2 \qquad y \times 4 = 20 \qquad \frac{10}{x} = 5$$

For $x + 3$, when you work out the value of any variables in the equation, it is called solving the equation.

Some equations can be solved by inspection. This means that you look at the equation and use reasoning to find the value of the variable.

Solve: $y + 4 = 11$

Reasoning: What number, added to 4 will give us 11?
              $4 + 7 = 11$, so y must be 7.

You can also use inverse operations to find the missing values.

Remember, if $7 + 4 = 11$, then $11 - 4 = 7$, so if $y + 4 = 11$, then $11 - 4 = y$.

The equals sign in an equation tells you that both sides of the equation have the same value. It also means that you can change the equation as long as you change the two sides in the same way and keep them equivalent.

Example

Solve: $\qquad x - 7 = 41$

You want to know what x is, so you have to get it on its own.

To do that, you can add 7

$x - 7 = 41$

$\downarrow \quad \downarrow$

$+ 7 + 7$

If you add 7 to the left-hand side, you need to add 7 to the right-hand side as well to keep the two sides equivalent.

You would show your working like this:

$x - 7 + 7 = 41 + 7$

$\qquad x = 48$

You can check the solution to an equation by putting the number you have found in place of the variable and making sure the number sentence is true.

$48 - 7 = 41$, so $x = 48$ is correct.

1 Solve these equations.

a $5 + x = 9$

b $12 + y = 29$

c $12 + y = 100$

d $11 + x = 50$

e $14 - x = 9$

f $15 - x = 1$

g $x + 8 = 23$

h $y + 35 = 50$

i $y + 1\frac{1}{2} = 4\frac{1}{2}$

j $3 \times x = 36$

k $x \times 9 = 36$

l $4 \times y = 24$

m $120 \div x = 40$

n $x \div 4 = 50$

o $350 \div y = 70$

2 Find the value of the variable in each equation.

a $a \times 6 = 48$

b $x + 44 = 52$

c $42 - x = 18$

d $x \div 2 = 9$

e $27 - a = 15$

f $m \times 5 = 40$

g $35 = s + 13$

h $15 \times m = 150$

i $32 \div z = 8$

3 Work out the value of *x* in each of these equations. Remember to apply the order of operation rules.

a $2 \times x + 10 = 25$

b $x \times 10 - 12 = 18$

c $(x + 10) \times 5 = 75$

d $3 \times x - 15 = 12$

e $5 \times x - 100 = 400$

f $x \times 5 + 14 = 44$

4 Find all the possible values of *x* and *y* in these equations.

a $140 \div 20 = x \times y$

b $40 - 20 = x + y$

c $40 - 10 = x \times y$

d $12 + 48 = 2 \times x + y$

Looking Back

Work with a partner to solve these equations.

a $11 = 9 + x$

b $4 \times y = 32$

c $(x - 12) \div 3 = 6$

d $16 - (x \div 3) = 4$

Unit 3 Use Equations to Solve Problems

Let's Think ...

The sum of an unknown number and 12 is 23.

- Write an equation to represent this problem.
- Solve the equation to find the value of the unknown number.

Expressions and equations are very useful as a problem solving strategy.

When you have a word problem you can choose variables (letters) to represent *unknown* amounts and write an equation to represent the problem.

It is important to say what the variable represents when you set out your work.

Read through these examples to see how to *create* equations and set out working clearly.

Example 1

The product of a number and 7 is 42. What is the number?

Let the number be x.	Say what the variable represents.
$7 \times x = 42$	Write an equation.
$x = 6$	Solve the equation. (Divide both sides by 7.)
$\therefore$ The number is 6.	Write an answer statement. $\therefore$ means therefore.

Example 2

If a number is multiplied by 2, and then 5 is added to the product, the result is 21. What is the number?

Let the number be y.	Say what the variable represents.
$(y \times 2) + 5 = 21$	Write an equation. This one has two steps.
$y \times 2 + 5 - 5 = 21 - 5$	Subtract 5 from both sides.
$y \times 2 = 16$	Solve. Divide both sides by 2.
$y = 8$	Check $8 \times 2 + 5 = 16 + 5 = 21$
$\therefore$ The number is 8.	Write an answer statement.

Example 3

After buying a packet of chips, Andy has 25¢ left from $1.00. How much did the chips cost?

Let the cost of the chips be p.
$1.00 = 100¢$.

$p + 25 = 100$	Write an equation. (This could also be $100 - p = 25$.)
$p + 25 - 25 = 100 - 25$	Solve the equation. Subtract 25 from both sides.
$p = 75$	Check: $75¢ + 25¢ = 100¢ = $1.00
The chips cost 75¢.	Write an answer statement.

1 For each problem, write an equation and solve it, even if you can see the answer without doing any working.

   a The sum of a number and 6 is 23. What is the number?

   b A number less 9 is 42. What is the number?

   c A number less 7 is 51. What is the number?

   d The product of a number and 8 is 72. What is the number?

   e When a number is divided by 5, the quotient is 8. What is the number?

   f Half of a number is added to 8 to get 23. What is the number?

2 Larry bought a pen and a sharpener for $2.25. If the pen cost $1.50, what did the sharpener cost?

3 It takes Jerome 45 minutes to do his mathematics and science homework. If the science took him 23 minutes, how long did the mathematics take?

4 Kezia bought tickets for a show for her and her friends. The tickets cost $5.00 each and the total paid was $105.00. How many tickets did she buy?

5 Nadia, Ann and Kim have a combined age of 27. If Nadia and Ann are the same age and Kim is 14, how old are Nadia and Ann?

6 Jerome put 156 counters into equal groups of 12. How many groups could he make?

7 The perimeter of each shape is given. Work out the length of each unknown side.

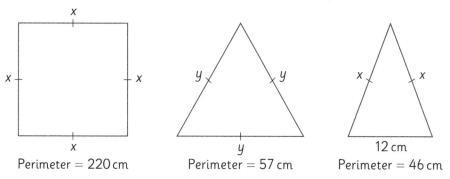

Perimeter = 220 cm   Perimeter = 57 cm   Perimeter = 46 cm

8 Tanya got 15¢ change from $1.00 when she bought 7 plums. How much did each plum cost?

9 A mango costs 10¢ more than an orange. Together a mango and an orange cost 98¢. What did each fruit cost?

10 Bruce weighs 6 kg less than Keshawn. If the sum of their masses is 88 kilograms, how much does each boy weigh?

Looking Back

What number must be added to 7 so that when the result is doubled, the answer is 22?

The length of a rectangle is 4 cm longer than its width. If the perimeter is 16 cm, how long are the sides?

Topic Review

Talking Mathematics

1 What mathematical operation symbol is correct for each word?

   a sum   b product   c quotient   d total

   e more than   f times   g minus   h less

2 How can you check whether the solution to an equation is correct or not?

Quick Check

1 Write an expression to represent:

   a the sum of a number and 23

   b the difference between a number and 19

   c the product of a number and 12

   d double a number

   e a number squared.

2 Write each of these as a mathematical equation and find the value of n.

   a The sum of n and 5 is 23.   b The sum of n and 9 is 17.

   c n less 12 results in 34.   d The product of n and 12 is 96.

   e The sum of half n and 13 is 63.

3 In how many different orders can you write the words 'good', 'better' and 'best'?

4 A carpenter plans to make 22 kitchen stools. Each stool can either have 3 or 4 legs. He has 81 legs in stock and he wants to use them all. How many of each type of stool can he make?

5 A number less 9 is equal to the product of 5 and 8. What is the number?

6 If a number is multiplied by 10, and then 6 is added to the result, the answer is 96. What is the number?

Topic 16 Ratio Workbook pages 45–48

Key Words

circle

circumference

compare

diameter

pi (π)

ratio

relationship

scale

▲ This photo of Earth was taken from space. Is there more water or more land on Earth? Which of these ratios do you think reflects the ratio of land to water: 1:2, 2:1 or 4:1?

We use **ratios** to **compare** numbers and amounts in everyday life. In this topic, you are going to learn more about how to use ratios to compare numbers and you are going to learn about a special number called **pi (π),** which we use to work out ratios in **circles**.

Getting Started

1 Can you explain what this instruction on a bottle of concentrate juice means?

   Dilute the concentrated juice as follows: 1 part of juice to 5 parts of water.

2 A healthy diet recommends that of each plate of food we eat:

   ● half should be vegetables and fruit

   ● a quarter should be proteins

   ● a quarter should be carbohydrates.

   Which of these statements show the amount of vegetables we should eat in relation to the rest of the food on the plate:

   $\frac{1}{2}$ 1:1   1:2   0.25   0.50

   vegetables and fruit

   protein

   carbohydrates

3 Think about your own diet. Draw a plate and divide it up like the one above to show what fraction of your entire day's food comes from each category.

Unit 1 Describing Ratios

Ratios can be used to compare two sets of data or numbers. Ratios, like fractions, show relationships between numbers, sets or parts of sets. You can write ratios and fractions in different ways.

Data: *There are 3 boys in a room. There are 9 girls in a room. The total number of people in the room is 12.*

Fractions: *A quarter ($\frac{1}{4}$ or 0.25 or 25%) of the people in the room are boys and three quarters ($\frac{3}{4}$ or 0.75 or 75%) of the people in the room are girls.*

Ratios: *The ratio of girls to boys in this room is nine to three ($9:3$).*
For every nine girls there are three boys.
The ratio of boys to girls in this room is three to nine ($3:9$).
For every three boys there are nine girls.

Be careful! Fractions and ratios may sometimes look similar but they are not the same.
$1:3$ and $\frac{1}{3}$ are not the same.

In the fraction ($\frac{1}{3}$), the denominator tells you the total number of parts in the whole (3).

In a ratio, you have to add up the two numbers to get the total, so $1:3$ means 1 out of a total of 4, which is the same as $\frac{1}{4}$.

Ratios do not have units. This is because the proportions are fixed, so the ratio works for any units.
A ratio of $1:2$ could mean 1 cup to 2 cups, 1 bag to 2 bags, 1 ton to 2 tons or any other appropriate units.
The order in which the quantities are written down in ratio is very important. Think about mixing orange juice from a concentrate. The instructions say 1 part concentrate to 3 parts water. This is a ratio of $1:3$.
This means that if you pour 1 cup of orange concentrate you have to add three cups of water. This will give you the correct mix and a nice drink.
You can reduce ratios to the lowest numbers to make equivalent ratios; for example:

> *$9:3$ can be reduced to $3:1$ by dividing both numbers by 3.*
>
> *$25:5$ can be reduced to $5:1$ by diving both numbers by 5.*

1   Read the following words. Write down the ratio of vowels to consonants in each word.

   a   ratio   b   rectangle

   c   parallelogram   d   mathematics

> The vowels are: a e i o u

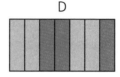

2 Write the ratio of the ingredients of each of the following mixtures.

a Two cups of flour, three cups of water.

b A kilogram of cement, 15 litres of water.

3 Look at the following shapes.

A   B   C   D

a What fraction of each shape is green?

b What is the ratio of green parts to red parts in each shape?

4 Look at the results of a survey conducted among students in a Grade 6 class. Say if each statement is TRUE or FALSE.

Morning Drink	
Juice	Tea
13	6

a The total number of people surveyed was 19.

b 6 out of 13 people drink tea in the morning.

c The ratio of people who drink tea to people who drink juice is 6:13.

5 Write the following ratios in their simplest forms (equivalent ratios).

a 4:6   b 75:100   c 20 cm:80 cm   d 0.4:0.8

6 Write equivalent ratios for the following sets of data.

a To make jam, add six kilograms of sugar for every three kilograms of fruit.

b To make brass you need to mix copper and zinc in the following ratio: 65:35.

c This spoon is made of pewter. 20% of the pewter is silver and 80% is tin.

d For every 15 minutes that I sit down, I spend 1 hour walking around.

7 Design a new flag for an Eco-club. The flag should have three colours. The club wants to promote care of the environment, which includes the land and the ocean. Choose three appropriate colours and design the flag. Write down the size of the flag and the ratios of the colours in the flag. Then use the ratios to calculate how much fabric in each colour you will need to make the flag.

8 A painter wanted to paint the walls of a room in pink. She mixed white and red paint in the ratio 4:1. She painted the wall but the colour was too light. Which of these ratios should she use to get a brighter shade of pink: 1:5 or 3:1?

9 You need to cook enough rice to feed a crowd of 36 people. The instructions on the packet say you need a ratio of one cup of rice to two cups of water to cook the rice successfully. The instructions also say that one cup of rice will feed four people. How many cups of water will you need?

Looking Back

Your teacher will give you a set of cards and some instructions to play a game using ratios.

Unit 2 Pi

Let's Think ...

Look at these three circles.

- Measure the circumference of each circle.
- Measure the diameter of each circle.
- What do you think the ratio between the circumference and the diameter of each circle is?

Pi is a special number for the ratio of the circumference of a circle to its diameter. You can find pi by dividing the circumference of a circle by its diameter. The ratio is always pi, no matter what the circumference of the circle is. The symbol for pi is π.

Pi is a very interesting number because it cannot be written down exactly. Its approximate value is 3.14 but the digits after the decimal point go on forever. You can also write pi as $\frac{22}{7}$.

Pi is also very useful; for example, if you know the diameter (d) of a circle, you can calculate its circumference (C) by using the formula $C = π × d$. If you know the radius (r), you can use this formula: $C = 2 × π × r$. Remember that the radius is exactly half of the diameter.

You can calculate the area (A) of a circle using this formula: $A = π × r^2$

Engineers and other scientist use pi in many calculations.

1 Work in pairs. Find six different objects with circles; for example, cups, bowls, clocks, pizzas. Measure the circumference of each circle. Measure the diameter. Then calculate the ratio (pi). Record your measurements and calculations on a table such as the one below, or use the table in your Workbook.

Item	Circumference	Diameter	Pi
Cup			

2 Work out the numbers that complete the sentences.

   a The radius of a circle with a diameter of 12 cm is ☐ cm.

   b The circumference of a circle is 22 cm. The diameter is ☐ cm.

   c The area of a circle with a radius of 8 cm is ☐ cm².

   d The ratio between the circumference of a circle and its diameter is approximately ☐ cm.

3 a Use a pair of compasses to draw two circles, one with a radius of 4 cm and the other with a radius of 6.5 cm.

   b Use a ruler to measure the diameter of each circle. Does it follow the rule that $d = 2r$?

   c Calculate the circumference of each circle using the formula $C = 2 × π × r$ or $C = π × d$.

   d Measure the circumference as accurately as you can with some string and compare your measurements with your calculations.

4 Find out what Pi Day is. Where and when do people celebrate Pi Day? What do they celebrate? How do they celebrate?

Looking Back

Explain to your partner what you have learned about pi and what you can use it for.

Unit 3 Scale

Let's Think ...

Look at the map.

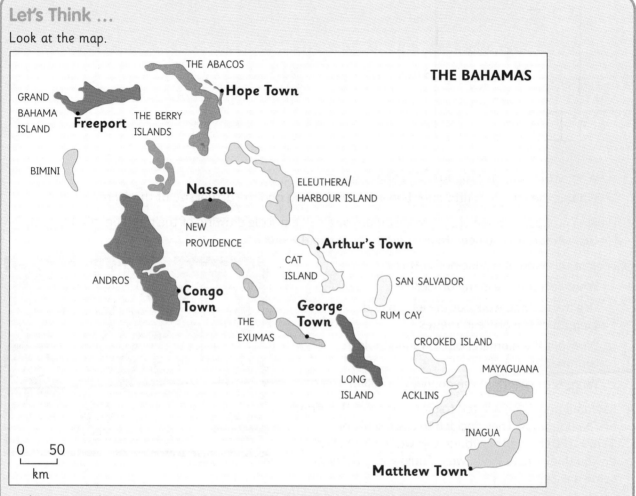

- Approximately how many kilometres would you have to travel to go from one end of Cat Island to the other?
- How can you work this out?

You learned about map scales in Topic 12. Remember that the scale given on a map gives the size of a map compared to the actual size of the places shown on the map. In other words, the scale shows the ratio between the sizes on the map and the real-life sizes.

A scale of 1:1 000 000 means 1 unit on the map = 1 000 000 units on the ground.

If 1 unit = 1 cm, then 1 unit on a map = 1 kilometre (1 000 000 cm) on the ground.

Most maps have a simple scale that allows you to measure and calculate distances easily. Here are two examples.

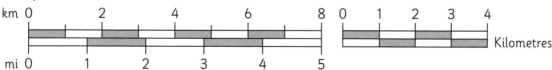

1 Study the following scale drawing of a room and answer the questions.

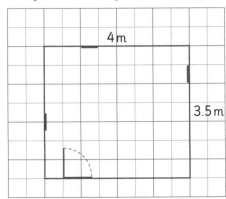

4 m

3.5 m

   a What scale was used to make this drawing?

   b What is the ratio in centimetres between the drawing and the actual size of the room?

2 Work in groups. Measure your own classroom and make a scale drawing of the room on graph paper. Discuss what ratio you need to use before you make the drawing.

3 Look at this plan of a principal's office.

   a What can you see in the office?

   b Use the scale to work out the length and breadth of the office in metres.

   d Work out the diameter of the round table in centimetres.

   d What is the width of the doorway in millimetres?

4 Seymour and Anne wanted to draw a map of the islands of The Bahamas. They decided to use a scale of 1 cm = 100 cm but they found this was difficult. What do you think the problem was? Could you give them some advice?

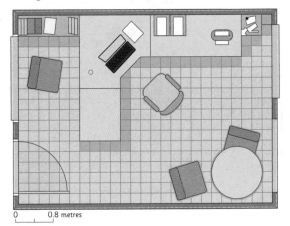

0   0.8 metres

5 Measure the following distances in millimetres as accurately as you can on the map of The Bahamas. Then, calculate the real distance using the scale information.

   a Length of Andros

   b Length of Bimini

   c Distance from Freeport to Hope Town

   d Distance between Congo Town and Matthew Town

   e Shortest distance from one end of The Exumas to the other.

6 Plot the shortest possible sailing route that includes a stop at every town shown on the map.

   a What distance would you sail?

   b Compare your route with a partner. Who has the shortest route?

Looking Back
Write a short explanation in your journal. Explain what scale is and why it is necessary on a map or plan.

Topic Review

What Did You Learn?

- You can compare numbers by using fractions, decimals, percentages or ratios.
- Ratios compare two sets of data or numbers, for example: $3:9$, $1:1000$.
- The fraction $\frac{2}{3}$ is **not** the same as the ratio $2:3$.
- Pi is a special number for the ratio of the circumference of a circle to its diameter.
- The approximate value of pi is 3.14 or $\frac{22}{7}$.
- The ratio of the circumference to the diameter of circle is always pi, no matter what the circumference of the circle is.
- The symbol for pi is π.
- You can calculate the circumference of a circle using the formulae: $C = \pi \times d$ or $C = 2 \times \pi \times r$.
- You can calculate the area of a circle using this formula: $A = \pi \times r \times r$.
- The scale of a drawing or a map shows the ratio between the sizes on the map and the real-life sizes.

Talking Mathematics

What is the mathematical word for each of these?

- The ratio that is true of all circles.
- The ratio used on maps to show how the size of the map relates to the size of a place on the ground.
- Two mathematical ways of comparing sets of data.

Quick Check

1 TRUE or FALSE?

    a The fraction $\frac{1}{2}$ can also be written as a ratio as $1:2$.

    b The ratio of consonants to vowels in the word SCALE is $3:2$.

    c The exact value of pi is 3.14.

    d π is the ratio of the diameter of a circle to its radius.

    e The ratio of the circumference to the diameter of a circle is the same for a circle with a circumference of 25 cm as for a circle with a circumference of 12 cm.

2 What is the real length represented by each line if it drawn to a scale of $1:120$

    a _____

    b _____

    c _____

3 Using a scale of $1:500$, draw a line to represent a real-life distance of 1 500 metres.

▲ A trip on the glass bottom boat in the photograph costs $15.00 per adult and $12.00 per child. How would you work out the cost for a group of 7 adults and 6 children? A group of school children go on the boat. The tickets cost $168.00. How would you work out how many children there were?

Earlier this year, you used mental methods to **multiply** and **divide** and you worked with factors and multiples. In this topic, you are going to revise some methods of multiplying and dividing larger numbers and use what you learn to solve problems, including problems with money amounts.

Getting Started

Talk about these two problems in groups.

a How do you know when a problem involves multiplication?

b How do you know when to divide to solve a problem?

c Which problem involves more than one step?

c Write number sentences and use them to solve each problem.

A tour operator buys tickets for the glass bottom boat for a group containing 12 adults and 8 children.

What is the total cost of the tickets?

The glass bottom boat operator does five trips on one day. The income from ticket sales is $6 840.00.

What is the average income per trip?

Key Words

multiply

product

divide

quotient

dividend

divisor

inverse

remainder

Unit 1 Use Mental Methods to Multiply and Divide

Let's Think …

- Explain how knowing that $6 \times 5 = 30$ can help you work out that $300 \div 6 = 50$.
- Given that $228 \div 12 = 19$, what is 12×19?
  How do you know that without doing any calculation?

Remember that *multiplication* and *division* are *inverse* operations.

You can use multiplication facts to work out division facts.

$12 \times 5 = 60$ So, $60 \div 5 = 12$ and $60 \div 12 = 5$

You can use division facts to work out multiplication facts.

$138 \div 3 = 46$ So, $3 \times 46 = 138$

Multiplying whole numbers by 10, 100, 1 000 or other multiples of 10 will give you zeros in the *product*. You can use your tables and what you know about place value to do this kind of multiplication mentally.

$9 \times 8 = 72$ $9 \times 80 = 720$ $90 \times 80 = 7\,200$

1 Calculate mentally. Write the answers only.

a 53×10	b 32×10	c 41×10
d 10×85	e 123×10	f 19×100
g 100×45	h 100×12	i 100×10
j $12 \times 10 \times 10$	k $23 \times 10 \times 100$	l $4 \times 10 \times 3$
m $3 \times 100 \times 2$	n $12 \times 100 \times 3$	o $10 \times 123 \times 100$

2 Divide mentally. Write the answers only.

a $420 \div 10$	b $876 \div 10$	c $990 \div 10$
d $4\,300 \div 10$	e $4\,300 \div 100$	f $2\,300 \div 100$
g $9\,000 \div 10$	h $32\,000 \div 100$	i $100\,000 \div 100$
j $3\,000 \div 100$	k $32\,000 \div 1\,000$	l $4\,200 \div 20$
m $540 \div 60$	n $320 \div 80$	o $1\,000 \div 50$

3 Work out the value of the letter in each equation by inspection.

a $x \times 14 = 28$	b $450 \div y = 5$	c $45\,100 \div m = 451$
d $1.85 \times n = 1\,850$	e $115 \div b = 23$	f $42 \times g = 2\,100$
g $480 \div x = 6$	h $y \div 30 = 6$	i $y \div 60 = 12$
j $101 \times m = 10\,100$	k $40 \times p = 2\,000$	l $2\,500 = m \times m$

Try to solve these problems using mental strategies if you can.

4   Sarai bought nine packs of crayons for $4.00 each. How much did she pay? How much change would she get from $50.00?

5   Joshua wants to give all his friends a popsicle. The popsicles are sold in packs of 4 and he needs 21. How many packs should he buy?

6   Mrs Smith has 50 roses. She wants to make ten bunches of 6. Does she have enough roses?

7   A tour operator needs to transport 28 people by cab. Each cab will take four people. How many cabs are needed?

8   Jayson runs 11 km every day. How far will he run in a week?

9   Malia bought six packs of hair grips. Each pack contained 4 hair grips and the total cost was $21.00.

    a How many hair grips did she buy?   b What was the cost of each pack?

10 Tonia is trying to get fit. She does 10 push-ups, 20 sit-ups and 5 star jumps every morning before school. At the weekend she does other exercises. How many of each exercise does she do in one week? How many does she do in total in a week?

11 A clothing factory uses 2 metres of cloth to make a pair of men's pants, 6 metres of cloth to make a tailored suit, 1.8 metres to make a pair of ladies pants and 0.8 metres to make a scarf.

    Work out how much cloth they would need to make:

    a 100 suits   b 1000 pairs of men's pants   c 100 scarves
    d 500 pairs of ladies pants   e 50 scarfs and 20 jackets

12 The factory has 800 m of printed fabric. What is the greatest number of ladies pants they can make from this?

13 Shawnae says:

    'You can make division easier if you halve both numbers; for example, 4 608 ÷ 18 is the same as 2 304 ÷ 9 and dividing by 9 is easier than dividing by 18.'

    a Check her reasoning using these four examples.
       8 208 ÷ 12   3 400 ÷ 16   9 156 ÷ 14   6 336 ÷ 18
    b Did halving the numbers work?
    c Did you find it an easier method?
    d Would it make all divisions easier? Explain your thinking.

Looking Back

Explain what happens when you multiply or divide a whole number by 1 000. Give examples to support your explanation.

Unit 2 Multiply and Divide Larger Numbers

Let's Think ...

- Paul says 174 ÷ 23 equals 6 remainder 35. How do you know he is wrong without checking the calculation?
- Kezia works out 124 × 20 and gets an answer of 248. What has she done wrong?

You learned how to multiply and divide larger numbers using pen and paper methods last year.

Remember to estimate before you calculate. Use the method that suits the numbers you are working with.

Read through the examples to help you remember how to do this.

Example 1

497×43

Long method

Estimate: $500 \times 40 = 20\,000$

$$
\begin{array}{r}
{}^{3}4\,{}^{2}9\,7 \\
\times \quad 4\,3 \\
\hline
{}^{1}1\,{}^{1}4\,9\,1 \\
{}^{1}1\,9\,8\,8\,0 \\
\hline
2\,1\,3\,7\,1
\end{array}
$$

× 43 is the same as (× 40) + (× 3)

← This is 3 × 497.

← If you multiply by 40 you will get one zero at the end of the answer. Write this in the ones place then work out 4 × 497.

Grid method

×	400	90	7
40	16000	3600	280	→ 19880
3	1200	270	21	→ 1491

$$21371$$

Example 2

How many times does 4 go into 784?

Estimate: $800 \div 4 = 200$

Short division

$$
\begin{array}{r}
1\,9\,6 \\
4\overline{)7\,{}^{3}8\,{}^{2}4}
\end{array}
$$

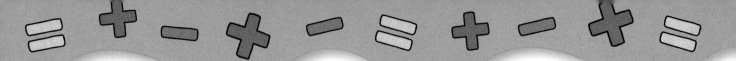

Example 3

What is $867 \div 21$?

Estimate: $860 \div 20 = 43$

Long division

```
      4 1 r 6
21| 8 6 7
   −84  ↓    21 × 4 = 84
      2 7    21 × 1 = 21
     −21
        6
```

The answer to a division is called the *quotient*. The number being divided is called the *dividend* and the number you are dividing into it is called the *divisor*.

When one number does not divide exactly into another, you are left with a *remainder*.

1 Estimate and then calculate using pen and paper methods. Check your answers using a calculator. If you have made a mistake, try to find it.

   a $3\,006 \times 8$ b $2\,406 \times 9$ c $3\,008 \times 7$ d 87×98 e 312×45

   f 423×87 g 145×208 h 876×39 i 412×602 j 342×456

   k 298×342 l 612×43 m $4\,123 \times 45$ n $1\,467 \times 234$ o $12\,098 \times 312$

2 Estimate and then calculate using pen and paper methods. Check your answers using a calculator. If you have made a mistake, try to find it.

   a $800 \div 7$ b $345 \div 6$ c $872 \div 8$ d $786 \div 24$

   e $832 \div 25$ f $347 \div 23$ g $1\,345 \div 21$ h $3\,214 \div 13$

   i $5\,330 \div 65$ j $12\,765 \div 27$ k $14\,235 \div 70$ l $10\,819 \div 51$

Hint

You can divide money amounts with decimal parts just like you divide whole numbers as long as you line up the decimal points in both the answer and the working part of the division.

3 The total cost for a number of people is given for each activity. Work out the cost per person.

   a Helicopter ride for 4 adults: $1245.80   b Hot air balloon trip for 5 people: $1241.75

   c Parasailing and snorkelling for 3 people: $218.97   d Deep sea fishing for six people: $2036.40

   e Waterpark and lunch for four people: $254.60

Looking Back

Estimate and then calculate. Show all your working.

   a 325×6 b 754×38 c 107×39

   d $98 \div 17$ e $130\,718 \div 6$ f $34\,213 \div 34$

Unit 3 Mixed Problems

When you solve word (or other) problems in mathematics, your first step is always to carefully read the problem and work out what you need to do to solve it.

You are expected to show how you solved each problem (your strategy) and to write down the steps you followed to work out the solution.

Showing your working is important. It allows you to keep track of what you have done and it also allows other people to follow your reasoning and see how you worked to solve the problem.

Read through the example to see how each of these students worked to solve the same problem.

Tonya has $20.00. She decides to buy some coloured pens with her money. The pens cost $2.95 each. How many pens can she buy? How much money will she have left if she buys as many pens as possible?

Method 1	Method 2
$2.95 \approx 3$	$^12^195$
$3 \times 6 = 18$	$+ \underline{2.95}$
$3 \times 7 = 21$	$^15.90$ for 2
She can buy 6 pens. ✓	$+ \underline{5.90}$ ← + 2 more
	$1^11.80$ for 4
$^5_2{}^395$	$\underline{5.90}$ ← + 2 more
$\underline{\times \quad 6}$	$^1^17.70$ for 6
$17.70 = $ cost	$\underline{2.95}$ ← +1 more
	20.65 Too many ✓
$^1{}_2{}^19{}^10.00$	She can buy 6 for $17.70.
$- \underline{17.70}$	
2.30	$20 - $17 = 3
She will have $3.30 left. ✓	$3 - $0.7 = 2.30 ✓
	She will have $2.30 left.

The actual method you use is generally up to you as long as you show how you got to the answer.

1. A school textbook has 128 pages. How many pages are in 502 of these books?

2. A factory makes 624 rum cakes each week. They close for two weeks over Christmas. How many cakes do they make per year?

3. A farmer has 1 276 pineapples. She wants to pack them into 58 crates and each crate should have the same number of pineapples. How many pineapples should be packed into a crate?

4. 460 students took part in a parade. They were placed in groups of 24 for the parade. How many groups of 24 could the organizer make? What do you think could be done with the remaining students?

5. Jan gets $10.00 per week for doing chores. She spends $8.25 of this each week and saves the rest. How long will it take her to save $25.00?

6. Karen bought 8 bottle of water costing $1.90 each and 8 cartons of juice costing $1.75 each. What is the total cost?

7. A car repair shop ordered 23 motors costing $257.00 each. What would the total cost be?

8. Claire's car uses a litre of gas per 12 kilometres on average. How many litres will it use if she travelled 850 kilometres?

9. There are 36 packets of chips in a box. Each packet costs 80 cents.
   a What does a box cost?
   b A school tuckshop sells 1 100 packs of chips per week. How many boxes should they order each week?
   c What would the cost of the weekly order be?
   d The tuckshops sells the chips for $1.00 per packet. How much profit do they make per week?

10. The grandstands at a sporting venue have 48 rows with 52 seats per row. How many seats is this?

11. A ship used 8 760 litres of fuel for 12 return journeys of the same distance. How many litres is this per single journey?

12. Ticket prices for a waterpark are displayed at the entrance:
    a What would the tickets cost for a group of 9 adults, 25 children and 32 students?
    b The total cost for a group of students is $896.00. How many were in the group?

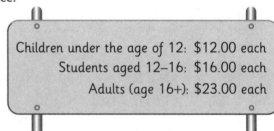

Children under the age of 12: $12.00 each
Students aged 12–16: $16.00 each
Adults (age 16+): $23.00 each

Looking Back
Make up three problems of your own. At least one of them must use money amounts.

Topic Review

What Did You Learn?

- Multiplication and division are inverse operations. You can use one fact to work out a related fact.
- To multiply larger numbers, you can use place value, a grid or a column method.
- Always estimate by rounding before you calculate.
- You can use short division as a written method to divide two-and three-digit numbers by a single digit.
- When you are dividing by a two-digit number, it is more efficient to record your work using long division.

Talking Mathematics

Explain how you would tackle this problem. (You do not need to solve it.)

A ship's engineer needs to saw a metal rod of 6 800 mm into 8 equal pieces. Cutting will waste 4 cm. How long will each section be?

Test your ideas using a calculator.

Quick Check

1 Tony has 2 462 bags of mangoes. He packs these into crates that can hold 32 bags. How many crates can he fill? How many bags of mangoes will be left over?

2 A rectangle is 14 times as long as it is wide. If it 23 cm wide, calculate its perimeter and its area.

3 The ferry from Nassau to Paradise Island costs $2.00. If there are 42 passengers per trip and the ferry crosses 11 times, how much money will they have collected in ticket fees?

4 How many 55 seater buses are needed to transport 857 people?

5 A regular hexagon has a perimeter of 3 438 cm. What is the length of each side?

6 Five businesswomen are sharing the cost of a meal in a restaurant. The total cost is $146.75. What is each person's share? They want to add at least 10 % to the bill as a tip, but they do not want to make the calculations too difficult. What would you suggest they do? Why?

7 Write a problem to match this number sentence.
$27.50 + (250 \times $3.00) = $777.50

Topic 18 Exploring Shape Workbook pages 53–58

▲ Can you identify the line of symmetry in each butterfly? Can you think of other animals and plants that are symmetrical?

A **line of symmetry** divides a shape into **congruent** shapes. Congruent shapes are exactly the same size and shape. When we cut out congruent shapes from paper or card, the shapes fit exactly on top of each other. In this topic, you will identify lines of symmetry and congruent shapes. You will also work with transformations: **flips**, **slides** and **turns**.

Getting Started

1 a How can you work out whether two shapes are congruent?

   b Which of the following shapes are congruent?

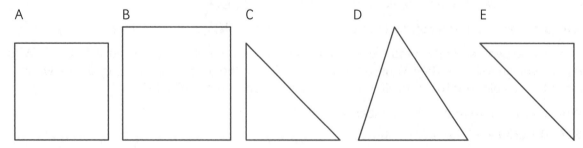

A   B   C   D   E

2 Work in pairs. You will need paper, a pencil, a ruler, and a pair of scissors.

   a Draw and cut out a square.

   b Now draw and cut out a rectangle.

   c How many ways can you fold each shape so that the two sides fit exactly on each other?

   d What are the differences and similarities between the ways you folded the rectangle and the square?

3 Draw a four-sided polygon that has no lines of symmetry.

Unit 1 Line Symmetry

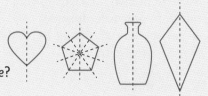
Line symmetry is also called reflection symmetry. A shape is symmetrical if you can draw a straight line that divides it into two congruent shapes. Remember, congruent shapes are exactly the same size and shape. A line of symmetry is a straight line that divides the shape into two exact mirror images. A shape can have more than one line of symmetry.

1 a Trace and cut out the following polygons.

b Fold the figures to work out whether they are symmetrical. If yes, how many lines of symmetry can you find?

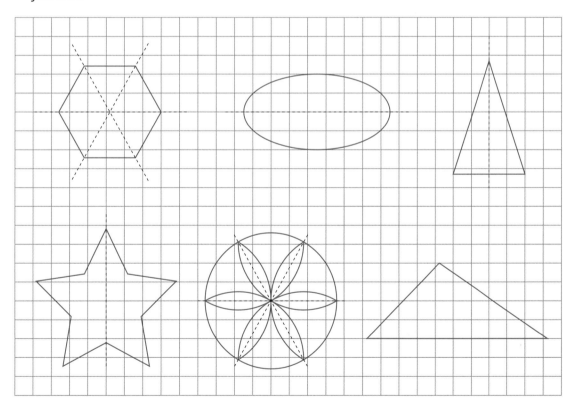

2 A MIRA is a tool you can use to investigate line symmetry in shapes. If you have a MIRA in your classroom, use it to show how each of your shapes is symmetrical.

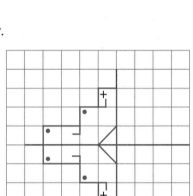

3 Make your own booklet showing the lines of symmetry of different polygons. Include:
  ● an equilateral triangle
  ● an isosceles triangle
  ● a square
  ● a rectangle
  ● a pentagon
  ● a hexagon
  ● an octagon
  ● at least three other shapes of your choice.

Use MIRAs to check that you have drawn your lines of symmetry correctly.

4 Half of a symmetrical pattern is given below.
  a Draw what the other half of the pattern will look like.
  b Use squared paper to draw half of a symmetrical pattern of your own.
  c Exchange with a partner and complete each other's patterns.

Looking Back

1 Look at these shapes from different religions, cultures and faiths. Identify the type of symmetry, and the lines of symmetry for each shape:

2 If two circles have the same radius length, are they congruent? Explain your answer.

Unit 2 Slides, Flips and Turns

Let's Think ...

How has the jigsaw piece been moved to get it from position A to position B in each case?

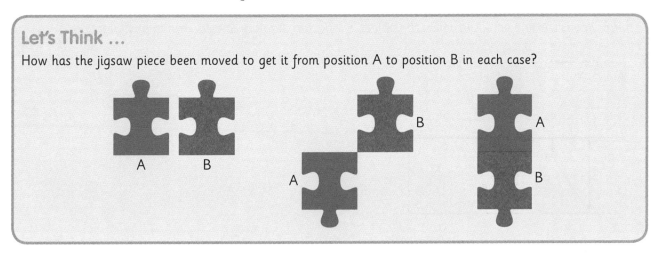

A transformation moves a polygon on a flat plane. It is also called a motion of transformation – in other words a movement that change's a shape's position in space. In this unit, you will show translations, reflections and rotations. The original shape is called the object. The new shape (or shape in its new position) is called the image.

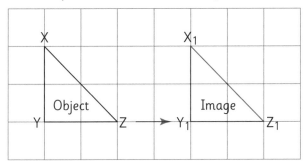

A translation (or slide) moves a shape in one direction. In this diagram, DEFG moves 4 units to the right:

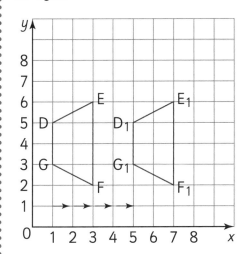

A *reflection* (or flip) moves a figure over a line, to create a mirror image. In this picture, ABC is reflected in the y-axis. When you reflect a shape, every point is the same distance away from the **mirror line**. The mirror line is the line or axis in which the object is reflected. A mirror line can run in any direction.

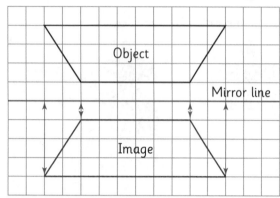

A *rotation* (or turn) is the movement of a figure around a vertex or point of rotation. When we describe a rotation, we say:

- the angle of the rotation (the measure of the turn in degrees)
- the direction of the rotation (clockwise or anticlockwise)
- the point about which the rotation turns (or the centre of rotation).

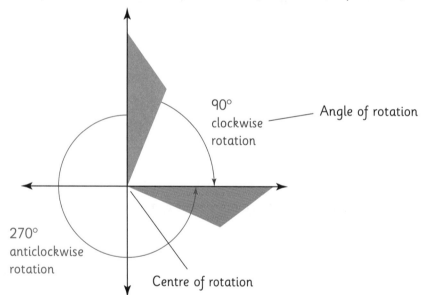

You can see that a 90° clockwise rotation produces the same image as a 270° anticlockwise rotation.

1 The diagram at the top of page 113 is a co-ordinate plane. It has an x-axis (horizontal axis) and a y-axis (vertical axis). We can describe points on the plane using co-ordinate pairs.

a Which axis runs from left to right (horizontally)?

b Which axis runs up and down (vertically)?

c If you draw a line to join points A and B, how many units long would the line be?

d If you draw lines to join A to B, B to C and C to A, what shape would you make?

e If you move the whole shape one unit to the right, what would the new co-ordinates be?

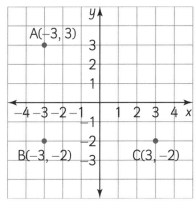

2 Look at each pattern below. To make each pattern, the artist took a single shape and changed its position.

   a Describe which shapes and movements the artist used.

   b Can you find a different way the artist could make the same pattern?

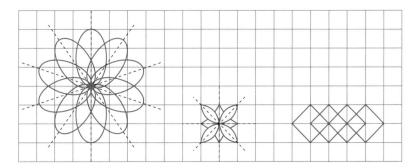

3 a Draw any quadrilateral on a piece of graph paper. Label it ABCD.

   b For each vertex (A, B, C and D), move 3 units up and 2 units to the right.

   c Mark the new corresponding vertices A_1, B_1, C_1 and D_1.

   d What do you notice about the two quadrilaterals?

4 a Write the co-ordinates of D, E, F and G from the object DEFG.

   b Write the new co-ordinates D_1, E_1, F_1, G_1 for the image DEFG.

   c If you translate the object DEFG down two units, what would the co-ordinates of the new image be?

5 Draw and cut out a polygon on some graph paper. On a fresh piece of graph paper, draw a set of co-ordinate axes (x-axis and y-axis). Trace your polygon to illustrate some different slides:

   a a slide from right to left along the x-axis

   b a slide that moves up the y-axis

   c a slide that moves up and left

   d a slide that moves down and right

6 a Write the co-ordinates of triangle ABC.

   b Write the co-ordinates of A_1, B_1, C_1 after it is reflected in the y-axis.

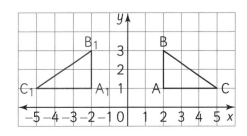

7 Draw and cut out a polygon. You are going to draw three different reflections of your shape on a piece of graph paper.

To draw a reflection of a given shape in a given mirror line, follow these steps:

- Measure the distance from each vertex to the mirror line, making sure you hold your ruler at a right angle (90°) to the mirror line.
- Measure the same distance again on the other side of the mirror line to each new vertex.
- Then join up the points you have drawn.

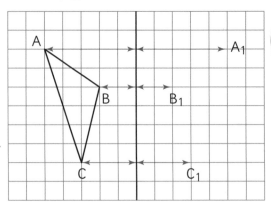

8 Try this challenge! With a partner, explore what happens to the co-ordinates of a polygon when you reflect it in:

a the *x*-axis

b the *y*-axis.

9 a Draw a rectangle 3 cm wide by 4 cm long.

b Trace your rectangle onto graph paper.

c Rotate your rectangle 90 degrees in a clockwise direction about one corner. Trace the shape in its new position.

d Repeat the rotation, tracing it each time, until your shape reaches its starting position.

e Compare shapes with others in your class. Do they look the same or different? Why?

10 Work on grid paper.

a Draw object ABC with A (5, 7), B (5, 11) and C (8, 11).

b Trace and cut out ABC.

c Rotate ABC 45° clockwise about point A. Trace the new shape. Write the co-ordinates of the new shape.

d Use your cutout to start with the original object. This time rotate it 90° clockwise about point C. Trace the new shape and write its co-ordinates.

11 Choose a polygon shape to use to create your own pattern. Use rotation to repeat your shape about a point to create a pattern of your own.

Looking Back

1 Copy this right-angled triangle. Perform the following transformations on it:

a Reflect the triangle in the side AB.

b Rotate the triangle 45° anticlockwise about the point B.

2 Draw a right-angled triangle on grid or graph paper. Translate your triangle four units up and two units to the right.

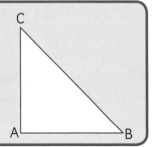

Topic Review

Talking Mathematics

1 The following words and phrases are all palindromes. A palindrome is a word, phrase, number or sequence which reads the same backwards and forwards:

A nut for a jar of tuna.

Madam I'm Adam.

I prefer pi.

Which geometry transformation does a palindrome remind you of? Why?

Can you think of any words or names that are palindromes? For a special challenge, come up with your own palindrome sentences!

2 Why do you think we need mathematical terms rather than everyday language?

Quick Check

1 What is a line of symmetry? Draw a diagram to explain your answer.

2 TRUE or FALSE? Rotations, translations and reflections change the size of shapes.

3 Match each type of transformation to the best description.

Type of Transformation	Description
Translation	Turns an object about a point
Rotation	Flips an object over a mirror line
Reflection	Moves an object up, down, left, right or diagonally along any straight line

4 Which type of transformation produces a mirror image of its object?

5 What angle of rotation would you use to turn a shape upside down?

Topic 19 Order of Operations Workbook pages 59–61

Key Words

operation

grouping symbols

brackets

multiply

divide

add

subtract

▲ This school hall is being remodelled. There are a lot of things that need to be done to completely remodel a room. Can you think of some of them? Some things have to be done before others. Can you think of some examples? What problems could result from doing things in the wrong order? Why?

Last year, you learned that when there is more than one **operation** in the same number sentence you have to do them in a particular order to avoid confusion and getting different answers. The order of operations rules are simple: work out **brackets** first, then **multiply** or **divide** working from left to right, then **add** or **subtract** working from left to right. These rules apply to all calculations including those with fractions and decimals. In this topic, you are going to revise the rules for ordering operations and then you are going to apply them to solve equations.

Getting Started

1 Quetel and Bonnie work out $12 \div 3 \times 2$. Quetel says the answer is 2; Bonnie says it is 8. Who is correct? Why?

2 Jerome has a mobile phone contract. His contract says he will pay 20¢ per minute for calls and 15¢ per text message. In the first month, Jerome sends 55 texts and makes 9 minutes of calls. This is how he works out what he needs to pay:

$9 \times 20 + 15 \times 55 = 10\,725$, which is $107.25

Jerome is shocked by this large amount. But he has made a mistake.

a What did he do wrong?   b What will he actually have to pay?

c How could you rewrite the calculation using brackets to show what parts need to be worked out first?

Unit 1 Revisit Order of Operations

Let's Think ...

What is $2 \times 5 + 6 \times 3$?

Try to write this calculation in three different ways that all give the same answer.

Do you remember the order of working with mixed *operations*?

The key rule of order of operations is that you always do operations inside *grouping symbols (brackets or parentheses)* first.

When there is more than one set of brackets in a number sentence, you work out the inside set first and then move outwards. Write down each step as you work.

- Next, *multiply* or *divide* in the order in which they appear (in other words, work from left to right).

- Then, *add* or *subtract* in the order in which they appear (again, work from left to right).

When you apply the rules, there is only one correct answer because there is only one correct order of working.

The examples show how to apply the order of operations correctly.

Example 1

$32 - 10 \times 2$ Multiply

$= 32 - 20$ Subtract

$= 12$

Example 2

$(14 + 1) \times 3 + 4$ Brackets

$= 15 \times 3 + 4$ Multiply

$= 45 + 4$ Add

$= 49$

Example 3

$40 \div 8 + 3 \times 7$ Divide Multiply

$= 5 + 21$ Add

$= 26$

1 Do these calculations correctly. Show your working.

  a $15 + 8 \times 3$ b $(15 + 8) \times 3$ c $30 - 4 \times 3$

  d $(30 - 4) \times 3$ e $32 \div 2 + 6$ f $32 \div (2 + 6)$

  g $49 - 21 \div 7$ h $(49 - 21) \div 7$

2 Work from the innermost to the outermost sets of brackets to correctly calculate these answers.

  a $3 + [19 - (2 \times 3)]$ b $25 - [5 + (16 \div 2)]$ c $[(36 \div 6) + 9] - 7$

  d $28 - [15 - (2 \times 6)]$ e $[4 \times (6 + 2)] \times 8$ f $[(21 \div 3) + 2] \times 9$

  g $28 \div [64 - (6 \times 10)]$ h $[90 \div (5 + 5)] \times 100$ i $3 \times [70 \div (22 \div 11)]$

  j $[(17 + 15) \times 2] \div 8$

3 Apply the order of operations rules to calculate these correctly.

  a $2 \times 7 + 2 \times 3$ b $2 + 7 \times 2 + 3$ c $4 \times 4 - 2 \times 2 \times 2$

  d $32 - 2 \times 8$ e $3 \times 4 + 4 \times 5 - 14 \div 2$ f $4 - 1 \times 2 + 3$

  g $(4 + 7) \times 3$ h $(17 - 9) \times 3 + 8$ i $30 - (6 - 2) \times 5$

  j $(4 + 4 \times 4) - 4$

4 Copy each equation and insert parentheses in the correct position to make the equation true.

  a $4 \times 5 + 7 = 48$ b $14 \div 10 - 3 = 2$ c $2 + 4 \times 3 - 7 = 11$

  d $7 \times 4 - 3 + 2 = 9$ e $12 \div 3 \times 2 = 2$ f $12 + 3 \times 0 = 0$

  g $8 + 6 \div 3 + 4 = 2$ h $100 - 3 \times 6 + 4 = 78$

5 Johnny was selling T-shirts at the market for $12.99 but he had to increase the price by $2.00 per T-shirt. A customer came to buy three T-shirts.

  Johnny entered this calculation into his calculator:

  $12.99 + 2 \times 3$

  a The calculator applied the rules of operations. What total would it give Johnny?

  b The new price of the T-shirts is $14.99. What is 14.99×3?

  c How could Johnny have used brackets to do the calculation correctly?

  d How much money would he have lost if he had charged the customer the wrong amount?

6 Add brackets to these calculations to get the highest possible answer.

  a $3 \times 2 \times 5 + 12$ b $7 - 3 \times 2 + 10$

7 Two adults and three children are going on a trip. The cost of the trip is $750.00 per adult and $450.00 per child. There is an extra charge of $75.00 for airport taxes.

  a Write an expression without brackets to represent the total cost.

  b Calculate the cost correctly.

Looking Back

Calculate correctly.

a $32 \times 3 + 34$ b $540 \div (423 - 413)$

c $350 \div 50 + 242$ d $12 \times (25 + 11) - 18$

Topic Review

What Did You Learn?

- In mathematics, you have to know and follow the rules to calculate correctly.
- Grouping symbols show you what operations to do first.
- Parentheses ( ), brackets [ ] and braces { } are all grouping symbols.
- When there are brackets inside brackets you work from the inside outwards.
- Operations are done in the following order: Brackets first → × and ÷ next (from left to right) → + and − last (from left to right).

Talking Mathematics

Describe in words the steps you would take to solve each of these calculations.

a $4 \times [60 \div (4 - 1)] + 25$ b $(15 + 6) \times 2 + (15 - 3 \times 2) - 16$

Explain why it is important to have rules such as the order of operation rules in mathematics.

Quick Check

1 Calculate.

a $(1 + 6) \times 20 \div 5$ b $15 + (4 \times 20) \div 5$ c $8 \times (4 \div 2) \times 3$

d $(8 \times 4) \div 2 \times 3$ e $10 + (5 - 2) \times 3$ f $10 + 5 - (2 \times 3)$

g $80 + 10 \div 10$ h $(80 + 10) \div 10$ i $11 \times 0 + 5$

j $11 \times (0 + 5)$ k $72 \div 6 \times (3 - 3)$ l $72 \div 6 \times 3 - 3$

2 Add brackets to each calculation to get the smallest answer possible.

a $12 \div 4 + 2 \times 3$ b $20 - 3 - 2 \times 4$ c $18 - 4 \times 2 - 3$

d $12 - 9 \times 24 - 22$ e $8 + 3 \times 30 \div 3 \div 11$ f $3 \times 4 - 2 \times 6$

3 Ms Rolle asked a group to calculate $6 \times 40 + (8 - 3)$. Only one student worked correctly. Who was it? What did the other two do incorrectly?

Rufus	Andrew	James
$6 \times 40 = 240$	$= 6 \times 40 + 5$	$= 6 \times 40 + 5$
$240 + 8 = 248$	$= 240 + 5$	$= 6 \times 45$
$248 - 3 = 245$	$= 245$	$= 270$

4 There are 240 students in a school. 2 groups of 12 leave school to take part in a science quiz and 5 groups of 6 leave school to take part in a mathematics quiz. Write and solve an equation to calculate how many students are left at the school.

Topic 20 Calculating with Fractions

Key Words

fraction

mixed number

equivalent

regroup

simplify

reduce

▲ Look at the photo. How can you describe the cut piece of apple mathematically? What is the mixed number that describes the whole group of apples? How many quarters are there in this amount of apples?

You already know how to find equivalent **fractions** and how to write fractions in simplest terms. Last year, you learned to add and subtract fractions and mixed numbers and how to regroup mixed numbers to make it easier to add and subtract them. In this topic, you are going to revise addition and subtraction of fractions and learn how to multiply fractions.

Getting Started

1 How many thirds are there in each of these?

    a $1\frac{1}{3}$ b $2\frac{2}{3}$ c $10\frac{1}{3}$

2 Delma has 11 lengths of ribbon. Each one is $\frac{1}{8}$ of a metre.

    a How many eighths is this altogether?

    b How many whole metres can she make? What fraction of a metre will she have left?

3 What is the value of x in each of these equations?

    a $\frac{4}{5} = \frac{x}{10}$ b $\frac{3}{5} = \frac{x}{20}$ c $1\frac{3}{4} = \frac{x}{4}$

4 What does it mean when you are told to simplify a fraction?

Unit 1 Add and Subtract Fractions and Mixed Numbers

Let's Think …

- How many metres of ribbon are there in all? $\frac{7}{12}$ m $\frac{1}{3}$ m $\frac{7}{9}$ m
- How much do the two bags weigh in total?

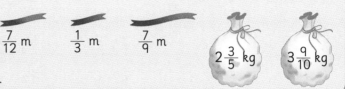

$2\frac{3}{5}$ kg $3\frac{9}{10}$ kg

Tell your partner how you worked out each total.

Do you remember how to add and subtract fractions and mixed numbers?

To add or subtract fractions, the denominators need to be the same.

When the denominators are different, you can convert them to equivalent fractions with the same denominator. You may need to regroup the answer if the numerator is greater than the denominator.

Example 1

$$\frac{3}{5} + \frac{9}{10} = \frac{6}{10} + \frac{9}{10} = \frac{15}{10}$$

$\frac{15}{10}$ means fifteen tenths. This is equivalent to 1 whole $\left(\frac{10}{10}\right)$ and $\frac{5}{10}$ or $1\frac{5}{10}$.

The $\frac{5}{10}$ part of the mixed number can be reduced to simplest terms to get $\frac{1}{2}$.

Example 2

$1\frac{11}{12} + 1\frac{7}{12}$

$1 + 1 = 2$ Add the whole numbers.

$\frac{11}{12} + \frac{7}{12} = \frac{18}{12}$ Add the fractions.

$\frac{18}{12} = \frac{12}{12} + \frac{6}{12} = 1\frac{1}{2}$

$2 + 1\frac{1}{2} = 3\frac{1}{2}$

Example 3

$1\frac{1}{3} + 1\frac{5}{6} - 2\frac{1}{4}$

$= \frac{4}{3} + \frac{11}{6} - \frac{9}{4}$ Regroup the mixed numbers.

$= \frac{16}{12} + \frac{22}{12} - \frac{27}{12}$ Convert to equivalent fractions with the same denominator.

$= \frac{38}{12} - \frac{27}{12}$ Work from left to right. Add: $\frac{16}{12} + \frac{22}{12} = \frac{38}{12}$

$= \frac{11}{12}$ Subtract next: $\frac{38}{12} - \frac{27}{12} = \frac{11}{12}$

1 Calculate. Give your answers in simplest terms.

a $\frac{3}{5}+\frac{1}{5}$ b $\frac{7}{5}+\frac{3}{5}$ c $\frac{7}{8}-\frac{5}{8}$ d $\frac{11}{5}-\frac{4}{5}$ e $3\frac{14}{16}-2\frac{11}{16}$ f $5\frac{9}{10}-5\frac{7}{10}$ g $3\frac{7}{8}-1\frac{7}{8}$

h $8\frac{3}{6}-3\frac{5}{6}$ i $1-\frac{3}{8}$ j $2-\frac{11}{15}$ k $4-\frac{5}{3}$ l $3-\frac{3}{4}$ m $2\frac{3}{8}-\frac{5}{8}$ n $3\frac{2}{7}-1\frac{5}{7}$

2 Calculate. Reduce your answers to simplest terms, if necessary.

a $\frac{3}{8}+\frac{3}{4}$ b $\frac{2}{3}+\frac{5}{6}$ c $\frac{3}{4}+\frac{9}{16}$ d $\frac{5}{8}+\frac{13}{16}$ e $\frac{9}{16}+\frac{1}{2}$ f $\frac{7}{10}+\frac{3}{5}$ g $\frac{7}{8}-\frac{3}{4}$

h $\frac{13}{16}-\frac{3}{4}$ i $\frac{3}{8}-\frac{1}{16}$ j $\frac{5}{7}-\frac{5}{14}$ k $\frac{17}{20}-\frac{3}{5}$ l $\frac{21}{25}-\frac{3}{5}$

3 Add. Give your answers in simplest form.

a $1\frac{3}{4}+2\frac{1}{4}$ b $1\frac{1}{5}+1\frac{2}{3}$ c $2\frac{1}{2}+3\frac{1}{4}$ d $1\frac{1}{8}+4\frac{1}{2}$ e $3\frac{7}{8}+\frac{5}{8}$ f $4\frac{3}{4}+2\frac{1}{2}$

g $2\frac{5}{8}+3\frac{1}{5}$ h $3\frac{7}{10}+1\frac{2}{3}$ i $2\frac{7}{50}+1\frac{19}{100}$ j $3\frac{3}{20}+2\frac{3}{4}$ k $3\frac{2}{3}+2\frac{4}{5}$ l $1\frac{2}{3}+3\frac{4}{5}$

4 Subtract. Give your answers in simplest form.

a $4\frac{8}{10}-1\frac{1}{2}$ b $3\frac{7}{8}-1\frac{1}{4}$ c $4\frac{9}{10}-2\frac{4}{5}$ d $3\frac{2}{3}-1\frac{1}{4}$ e $5\frac{1}{2}-3\frac{7}{8}$ f $4\frac{3}{5}-1\frac{9}{10}$

g $5\frac{1}{4}-2\frac{5}{12}$ h $4\frac{1}{2}-2\frac{2}{3}$ i $4\frac{3}{10}-2\frac{2}{3}$ j $4\frac{3}{8}-1\frac{2}{5}$ k $1\frac{3}{8}-\frac{11}{16}$ l $7\frac{7}{15}-4\frac{11}{12}$

5 Calculate.

a $\frac{3}{10}+\frac{2}{5}+\frac{4}{15}$ b $\frac{3}{4}+\frac{5}{6}+\frac{1}{2}$ c $\frac{3}{4}-\frac{1}{3}-\frac{1}{6}$ d $\frac{3}{4}+\frac{2}{5}-\frac{7}{10}$ e $\frac{2}{3}+\frac{3}{4}-\frac{1}{5}$

f $\frac{3}{4}-\frac{1}{3}+\frac{1}{2}$ g $2\frac{5}{9}+2\frac{5}{6}-2\frac{5}{18}$ h $4\frac{3}{4}-1\frac{2}{3}+3\frac{1}{2}$ i $2\frac{1}{4}+3\frac{1}{6}-3\frac{1}{8}$ j $3\frac{1}{5}+2\frac{1}{15}-2\frac{3}{4}$

6 Try to solve these equations. Show your working.

a $x-4\frac{1}{2}=1\frac{1}{6}$ b $3\frac{1}{4}+x=5\frac{19}{20}$ c $x-2\frac{1}{4}=3\frac{3}{5}$

7 Mrs Newton is baking. She mixed together $3\frac{1}{2}$ cups of brown flour, $1\frac{2}{3}$ cups white flour, $\frac{2}{3}$ of a cup of bran and $\frac{1}{8}$ of a cup of sugar. How many cups is this altogether?

8 Jeanne swam $12\frac{1}{3}$ lengths of the pool and Keishla swam $14\frac{3}{4}$ lengths of the pool. How much further did Keishla swim?

9 What is the perimeter of a rectangular lawn measuring $12\frac{1}{4}$ m by $4\frac{4}{5}$ m?

Looking Back

Calculate

a $4-2\frac{3}{5}$ b $3\frac{3}{4}-2\frac{1}{10}$ c $5\frac{1}{3}+2\frac{6}{7}$

Unit 2 Multiply Fractions

Let's Think …

- Sandra drank $\frac{1}{5}$ of a litre of juice every day for four days. How much juice is this altogether?

- Sherrie has three bars of chocolate. She gives $\frac{2}{5}$ of the chocolate to her sister. How much is $\frac{2}{5}$ of three bars?

$\frac{1}{5}$ L

Chocobar

Chocobar

Chocobar

You can solve the problems in the Let's Think … activity by adding the fractions.

$$\frac{1}{5} + \frac{1}{5} + \frac{1}{5} + \frac{1}{5} = \frac{4}{5}$$

$$\frac{2}{5} + \frac{2}{5} + \frac{2}{5} = \frac{6}{5} = 1\frac{1}{5}$$

Repeated addition can also be done as a multiplication.

Remember, any whole number can be written as a fraction with a denominator of 1.

$4 \times \frac{1}{5}$

$= \frac{4}{1} \times \frac{1}{5}$

$= \frac{4 \times 1}{1 \times 5}$ *Multiply numerators by numerators and denominators by denominators.*

$= \frac{4}{5}$

$3 \times \frac{2}{5}$

$= \frac{3}{1} \times \frac{2}{5} = \frac{3 \times 2}{1 \times 5} = \frac{6}{5}$

$= 1\frac{1}{5}$

Example 1

What is $\frac{3}{10}$ of 40?   The word 'of' in a mathematical sense, means multiply.

$$\frac{3}{10} \times \frac{40}{1} = \frac{3 \times 40}{10 \times 1} = \frac{120}{10} = 12$$

You can use the same methods to multiply fractions by other fractions.

Example 2

What is $\frac{3}{4}$ of $\frac{1}{2}$?

$$\frac{3}{4} \times \frac{1}{2} = \frac{3 \times 1}{4 \times 2} = \frac{3}{8}$$

Look at the diagram to check that this is correct.

1 whole

$\frac{1}{2}$

$\frac{3}{4}$ of $\frac{1}{2}$

Which is $\frac{3}{8}$ of the whole.

You can make multiplication of fractions easier for yourself if you *simplify* before you multiply. You will have smaller numbers to work with and you will get the same answer.

Look at these two methods.

Method 1: Simplify first.

$\frac{1}{3} \times \frac{9}{10}$

$= \frac{1}{3^1} \times \frac{\overset{3}{\cancel{9}}}{10}$ $\left(\text{this is the same as } \overset{\div 3}{\div 3}\right)$

$= \frac{3}{10}$

Method 2: Simplify the answer.

$\frac{1}{3} \times \frac{9}{10}$

$= \frac{9}{10}$ Divide by $\frac{3}{3}$ to simplify

$= \frac{3}{10}$

The way of showing the working in Method 1 is called *cancelling*.

Remember that to keep the fractions equivalent, you have to do the same to the numerator and denominator. So you can only cancel by dividing a numerator and denominator by the same amount.

To multiply mixed numbers, you must regroup them and write them as improper fractions before you multiply numerators by numerators and denominators by denominators.

1   Calculate.

a $\frac{1}{2} \times \frac{3}{4}$ b $\frac{3}{4} \times \frac{5}{6}$ c $\frac{7}{8} \times \frac{7}{8}$ d $\frac{3}{5} \times \frac{2}{3}$ e $\frac{2}{3}$ of $\frac{1}{6}$ f $\frac{5}{6}$ of $\frac{3}{10}$ g $\frac{5}{8}$ of 16   h $\frac{5}{6}$ of $\frac{5}{8}$

2   Multiply. Simplify before you multiply (cancel) where possible.

a $\frac{8}{9} \times \frac{3}{5}$ b $\frac{7}{10} \times \frac{5}{8}$ c $\frac{1}{3} \times \frac{9}{20}$ d $\frac{1}{2} \times \frac{4}{5}$ e $\frac{1}{2} \times \frac{8}{10}$ f $\frac{3}{10} \times \frac{5}{6}$

g $\frac{5}{8} \times \frac{2}{10}$ h $\frac{2}{3} \times \frac{3}{4}$ i $3 \times \frac{5}{6}$ j $\frac{11}{12} \times 4$ k $\frac{3}{5} \times \frac{10}{13}$ l $\frac{7}{10} \times \frac{5}{21}$

3   Find the product.

a $\frac{3}{4} \times \frac{9}{10}$ b $\frac{15}{20} \times \frac{3}{4}$ c $\frac{18}{25} \times \frac{3}{4}$ d $\frac{3}{10} \times \frac{15}{16}$ e $\frac{49}{100} \times \frac{3}{7}$ f $1\frac{1}{3} \times \frac{3}{4}$

4   Apply the correct order of operations rules to do these mixed calculations.

a $\frac{1}{2} \times \frac{3}{5} - \frac{1}{4}$ b $\left(\frac{3}{8} + \frac{1}{3}\right) \times \frac{3}{10}$ c $\frac{1}{4} \times \frac{2}{3} + \frac{2}{3}$ d $\frac{1}{3} \times \frac{1}{4} + \frac{3}{8}$ e $\frac{5}{9} - \frac{3}{10} \times \frac{5}{6}$

f $\frac{3}{5} + \frac{7}{10} - \frac{1}{2} \times \frac{4}{15}$ g $\frac{14}{15} \times \frac{5}{7} + \frac{4}{5}$ h $\left(\frac{2}{3} + \frac{5}{6}\right) \times \frac{1}{2}$ i $\frac{3}{4} - \frac{2}{5} \times \frac{1}{2}$

5   Darren has to walk $\frac{9}{10}$ of a kilometre to school. He is $\frac{1}{3}$ of the way there. How much further does he have to walk?

6   Mr Benson has 1 740 mangoes to sell. He has sold $\frac{5}{12}$ of them. How many has he sold?

7   $\frac{4}{5}$ of 320 students walk to school. How many do not walk?

8   What is the area of a rectangular flower bed that is $\frac{3}{5}$m wide and $\frac{7}{8}$m long?

Looking Back

Calculate.

a $\frac{2}{3}$ of 21   b $\frac{3}{4} \times \frac{1}{5}$ c $\frac{12}{16} \times \frac{24}{30}$ d $\frac{2}{3} \times \left(\frac{5}{8} - \frac{1}{4}\right)$ e $\frac{5}{6} \times \frac{21}{25} + 1\frac{1}{5}$

Topic Review

Talking Mathematics

- What did you find easiest in this topic? Why?
- What did you find most challenging? Why?
- What three things are important when you work with fractions? Why?

Quick Check

1 Do these calculations as quickly and efficiently as you can.

a $\frac{4}{5} + \frac{1}{3}$

b $2\frac{2}{5} + \frac{3}{10}$

c $\frac{3}{4} - \frac{3}{8}$

d $2\frac{3}{4} - 1\frac{7}{8}$

e $\frac{2}{3}$ of 60

f $\frac{3}{4}$ of $24

g $\frac{3}{4}$ of $10

h $\frac{7}{8} \times \frac{3}{5}$

i $\frac{2}{3} \times \frac{5}{8}$

j $\frac{5}{3} - \frac{3}{4} + \frac{1}{6}$

k $\frac{5}{3} - \left(\frac{3}{4} + \frac{1}{6}\right)$

l $\frac{2}{3} + \frac{1}{3} \times \frac{1}{4}$

m $\frac{2}{3} \times \frac{1}{2} + \frac{3}{7}$

n $\left(\frac{2}{3} + \frac{5}{8}\right) \times \frac{1}{2}$

o $\frac{3}{4} \times \frac{5}{8} - \frac{1}{2} \times \frac{1}{8}$

2 Mira carried three bags of shopping home from the supermarket for her gran. The bags weighed $2\frac{1}{10}$ kg, $3\frac{3}{4}$ kg and $2\frac{4}{5}$ kg. What was the total mass of the bags?

3 $1\frac{4}{5}$ litres of water is poured from a $2\frac{1}{2}$ litre container. How much water is left?

4 Make up three multi-step addition and subtraction problems involving fractions. Swap with a partner and solve each other's problems.

5 Mr Butler is a keen gardener. He used $\frac{1}{8}$ of his garden for vegetables, $\frac{1}{2}$ for lawn and $\frac{1}{3}$ for fruit trees. What fraction of the garden is left for growing flowers?

Topic 21 Calculating with Decimals Workbook pages 64–68

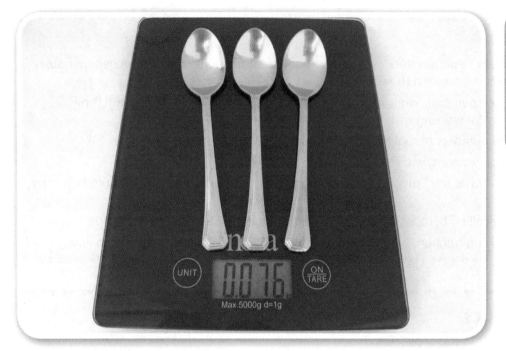

▲ Read the mass of these three items on the scale display. What operation would you need to do to find the mass of one item? What would the scale display if another three of these items were added to it?

We use decimals in many different situations in our everyday lives so it is important that you know how to calculate with decimals. Last year, you learned to add and subtract decimals using place value and you have worked with decimals to solve measurement and money problems. In this topic, you are going to extend your skills and learn to multiply and divide decimals. You will then practice and apply what you have learned.

Getting Started

1 Ellsworth says adding and subtracting decimals is easy as long you remember the PUP rule – Points Under Points.

  a What does he mean when he says this?

  b What do you do with empty spaces in columns when you add or subtract decimals?

  c What is $14.5 + 12.4$?

  d What is $14.5 - 12.4$?

2 Make a list of places where you might need to calculate decimals in everyday life. Try to find at least ten different examples. Share your ideas with the class.

Unit 1 Add and Subtract Decimals

Let's Think …

The odometer in Ms Giffard's car showed 321.9 at the start of a journey.

What does that mean?

At the end of her journey, the odometer showed 537.6.

How far did she travel?

To add or subtract decimals, make sure the *decimal points* are aligned below each other, then add or subtract as you would with whole numbers. Empty places may be filled with zeros.

Estimate by rounding before you calculate.

Calculate:

a $12.045 + 2.3 + 0.8 + 6$

Estimate: $12 + 2 + 1 + 6 = 21$

```
        ┌─── Line up decimal points
        ↓
  12.045
   2.300
   0.800 ←── Write 0 in each empty place
 + 6.000 ←── Remember 6 = 6.0
  ───────
  21.144
  11↶ Regroup 11/10 = 1.1
```

$Regroup \dfrac{11}{10} = 1.1$

b $23.8 - 9.03$

Estimate: $24 - 9 = 15$

```
  1 1 7 1         ┌─── Line up decimal points
  2̶3̶.8̶0 ←── Write 0 in the
                  empty place
 ↗−09.03
  ──────── ←── Rename as you
   14.77          need to
```

You can put 0 before whole numbers too if it helps

---

1 Try to do these using mental methods. Write the answers only.

a $0.4 + 0.2$	b $0.8 + 0.1$	c $0.4 + 0.5$	d $0.1 + 0.1$
e $0.5 + 0.6$	f $0.2 + 0.23$	g $0.23 + 0.1$	h $0.9 + 0.3$
i $0.8 - 0.3$	j $0.9 - 0.6$	k $0.8 - 0.1$	l $0.7 - 0.6$
m $2 - 0.5$	n $2.4 - 1.2$	o $16.5 - 4.5$	p $6.9 - 5.2$

2 Write in columns and calculate.

   a 0.23 + 0.93   b 0.37 + 0.65   c 0.42 + 0.55   d 0.28 + 0.67

   e 1.49 + 0.99   f 2.34 + 0.07   g 4.09 + 2.8   h 5.32 + 5.48

   i 0.87 − 0.43   j 0.9 − 0.65   k 0.86 − 0.4   l 0.4 − 0.23

   m 14.3 − 3.09   n 12.09 − 4.5   o 5.45 − 0.99   p 3.67 − 2.8

   q 16 − 5.234   r 9 − 1.008   s 25 − 14.809   t 342 − 45.09

3 Estimate and then calculate.

   a 5.99 + 15.32 + 231.09   b 214.6 + 87.99 + 234.9   c 612.5 + 132.09 + 99.5

   d 231.9 + 54.3 + 9.085   e 325.3 − 124.865   f 243 − 124.55

   g 412.89 − 128.805   h 100 − 45.087

4 Estimate and then calculate these decimal quantities. Remember to include units in your answers.

   a $5.87 + $500 + $235.50 + $100.99   b 6.876 km + 500 km + 1200 m (convert to km first!)

   c 56.004 seconds + 12.5 seconds   d 78.13 kg − 32.95 kg

   e 300 litres − 234.565 litres   f 12 cm + 123 mm + 3.4 cm + 0.9 cm (convert to cm first!)

5 A shopkeeper buys 50 m of rope. He sells 13.5 m and 8.25 m. How much is left?

6 Percy is on a diet. He loses 0.75 kg one week and 1.08 kg the next week. How much does he lose in total?

7 Debbie gets a bill for $28.75 and pays with two twenty dollar bills. How much change will she receive?

8 A small plane flies 123.65 km on Friday, 132.08 km on Saturday and 109 km on Sunday. How far is this altogether?

9 Shamila has a mobile phone app that maps her route when she goes for a ride on her bicycle. The distance for each sector is shown. Look at the route map and work out how far Shamila cycled.

10 Passengers are allowed to carry hand baggage onto an airplane as long as their luggage weighs less than 7 kg. Juan has a bag that weighs 1.2 kg. He packs a laptop weighing 2.45 kg and some books with a mass of 1.5 kg into the bag. His mom gives him a gift for his aunt that weighs 0.459 kg and a 900 gram packet of chips. He also has a bottle of water that weighs 250 grams. Will his bag be allowed on board?

Looking Back

1 Calculate.

   a $32.65 + $23.08 − $12.00

   b 6.51 kg + 14 kg + 12.8 kg

2 In New York, the temperature at 6:00 a.m. was 12.8 °C. By noon, it was 4.8 degrees warmer and it increased by a further 2 degrees in the afternoon. By early evening, the temperature dropped 2.9 degrees and it continued to decrease another 8.4 degrees to midnight. What was the temperature at midnight?

Unit 2 Multiply and Divide Decimals by Powers of 10

Let's Think ...

Try to do these calculations mentally.

8.56 × 10	8.56 ÷ 10
8.56 × 100	8.56 ÷ 100
8.56 × 1000	8.56 ÷ 1000

What is the short method or mental strategy for multiplying and dividing by 10, 100 or 1000?

Do you remember how to multiply and divide by powers of 10 mentally?

When you multiply by 10, each digit moves one place to the left to make the answer greater.

$15 \times 10 = 150$

$1.5 \times 10 = 15$

$1.05 \times 10 = 10.5$

Division is the inverse, or opposite, of multiplication, so when you divide by 10, the digits move one place to the right, making the answer smaller.

$150 \div 10 = 15$ *(Can you see that this is the same as 15.0?)*

$15 \div 10 = 1.5$

$1.05 \div 10 = 0.105$

When you multiply or divide by 100 or 1000, you move digits one place for each power of ten. So for 100 you move two places and for 1000 you move three places. The direction of the movement depends on whether you are multiplying or dividing.

$12 \times 10 = 120$	$1.2 \times 10 = 12$	
$12 \times 100 = 1200$	$1.2 \times 100 = 120$	
$12 \times 1000 = 12000$	$1.2 \times 1000 = 1200$	
$1200 \div 10 = 120$	$12 \div 10 = 1.2$	
$1200 \div 100 = 12$	$12 \div 100 = 0.12$	Think $\dfrac{12}{100} = 0.12$
$1200 \div 1000 = 1.2$	$12 \div 1000 = 0.012$	Think $\dfrac{12}{1000} = 0.012$

---

1 Write the power of ten represented by each variable in each set of operations.

Set 1

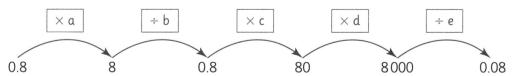

Set 2

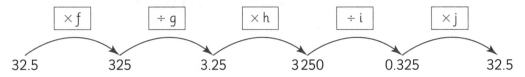

2 Copy and complete each of these sequences by filling in the missing operations.

a

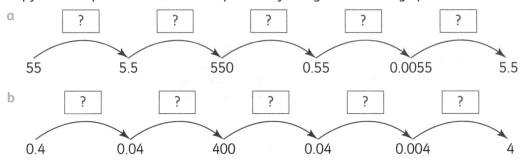

b

3 Multiply.

a 2.34×10	b 32.5×10	c 0.45×10	d 0.08×10
e 54.34×10	f 0.008×10	g 1.2×10	h 43.2×10
i 4.56×100	j 32.45×100	k 9.45×100	l 9.321×100
m 0.08×100	n 4.32×1000	o 1.2×1000	p 0.8×1000
q 7.99×100	r 6.5×1000	s 0.008×10	t 0.075×1000

4 Divide.

a $21.5 \div 10$	b $143 \div 10$	c $1.65 \div 10$	d $32.5 \div 10$
e $1324.5 \div 10$	f $54.7 \div 10$	g $0.16 \div 10$	h $0.87 \div 10$
i $132 \div 100$	j $55 \div 100$	k $67 \div 100$	l $67 \div 1000$
m $560 \div 1000$	n $56 \div 1000$	o $543.6 \div 1000$	p $1566 \div 1000$

5 A builder requires 100 m of timber at \$12.68 per metre. What will it cost him?

6 The overall cost of a wedding reception for 100 guests was \$12 345.50. What was the cost per guest?

7 Ten thousand people each paid \$1.99 for a charity raffle ticket.

a How much money was raised through ticket sales?

b The money from ticket sales was shared evenly between 100 different children's charities. How much did each receive?

Looking Back

Try to do these mentally. Write the answers only.

a 3.6×10	b 6.5×10	c 0.4×10	d 1.2×10
e $0.2 \div 10$	f 1.23×100	g $456 \div 100$	h 0.008×100
i $0.04 \div 10$	j $23.5 \div 100$	k $16.2 \div 100$	l $2.5 \div 1000$

Unit 3 More Multiplying and Dividing with Decimals

Let's Think ...

- What is the total length of these four pieces of ribbon?
- Write this problem as a multiplication number sentence.
- How many decimal places are in the question? And the product?

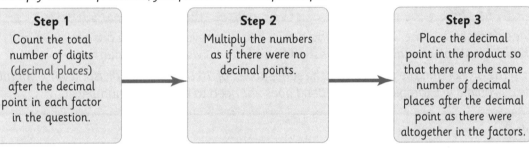

0.12 m   0.12 m

0.12 m   0.12 m

To multiply decimal fractions, you follow three simple steps.

Step 1	**Step 2**	**Step 3**
Count the total number of digits (decimal places) after the decimal point in each factor in the question.	Multiply the numbers as if there were no decimal points.	Place the decimal point in the product so that there are the same number of decimal places after the decimal point as there were altogether in the factors.

Example 1

a *5 × 3.43*

5 × 3.<u>43</u>   Two decimal places in the factors.

$^{2}\,^{1}$
3 4 3
× 5
———
1715

5 × 3.43 = 17.15

↑

Insert decimal point so product has two decimal places.

b *0.7 × 0.8*

0.<u>7</u> × 0.<u>8</u>   Two decimal places in the factors.

7 × 8 = 56   0.75 × 0.8 = 0.<u>56</u>

↑

Insert decimal point so product has two decimal places. Write 0 to show there are no whole numbers.

c *3.25 × 4.8*

3.<u>25</u> × 4.<u>8</u>   Three d.p. in the factors.

325
× 48
———
2 600
13 000
———
15 600

3.25 × 4.8 = 15.<u>600</u>

↑

Insert decimal point so there are three decimal places in the product.

In calculation c you can write the product as 15.6, but only do this after you have inserted the decimal point correctly.

When you divide a decimal by a whole number, place the decimal point in the quotient above the decimal point in the dividend. This works for both short and long division.

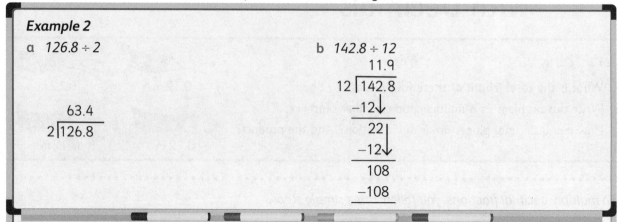

Example 2

a $126.8 \div 2$

$$\frac{63.4}{2\overline{)126.8}}$$

b $142.8 \div 12$

$$\begin{array}{r} 11.9 \\ 12\overline{)142.8} \\ -12\downarrow \\ \overline{22} \\ -12\downarrow \\ \overline{108} \\ -108 \end{array}$$

You can divide decimals by other decimals using a calculator. However, when you are using pen and paper methods, it is simpler to change the question to make the divisor a whole number.

Multiply both the divisor and the dividend by the same power of 10 to make the divisor a whole number. Then, divide as you did in Example 2. The dividend does not need to be a whole number.

Example 3

a $2.6 \div 0.2$ (Multiply both values by 10 to change 0.2 to a whole number.)

$$2.6 \div 0.2 = 26 \div 2$$
$$\times 10 \quad \times 10$$
$$= 13$$

b $2.65 \div 0.5$ (Multiply both values by 10 to change 0.5 to a whole number.)

$$2.65 \div 0.5 = 26.5 \div 5$$
$$\times 10 \quad \times 10$$

$$\begin{array}{r} 5.3 \\ 5\overline{)26.5} \\ 25 \\ \overline{15} \\ 15 \end{array}$$

1 Calculate.

a 4.5×4	b 2.4×8	c 3.7×5	d 3.23×3	e 1.25×6
f 3.42×9	g 3.45×12	h 4.56×20	i 25.4×3.1	j 2.8×2.5
k 9.13×2.5	l 0.89×0.4	m 8.7×300	n 0.76×0.23	o 1.24×8.25

2 Calculate.

a $0.9 \div 3$	b $0.8 \div 4$	c $1.2 \div 4$	d $4.9 \div 7$	e $6.4 \div 8$
f $0.36 \div 2$	g $3.6 \div 0.3$	h $0.24 \div 0.4$	i $0.8 \div 0.02$	j $5.25 \div 0.5$
k $4.5 \div 0.15$	l $9.72 \div 2.7$	m $18.81 \div 0.09$	n $2.718 \div 0.03$	o $25.5 \div 0.05$

Looking Back

Rewrite these calculations with the decimal point in the correct position.

a $1.2 \times 0.9 = 108$ b $1.9 \times 2 = 38$ c $0.8 \times 0.8 = 64$

Topic Review

Talking Mathematics

Discuss these questions in groups.

How can you use estimation to decide whether your answer to a calculation involving decimals is reasonable or not?	Why is estimation important when you use a calculator to work with decimals?	Do the expressions $1.8 \div 2$ and $0.18 \div 0.2$ have the same value? Explain your answer.	Is it true that you can just ignore decimal points when you are multiplying decimals? Explain your answer.

Quick Check

1 Calculate.
   a $4.7 + 12.65 + 0.812 + 12$
   b $18.8 - 9.25$ c $2 - 0.875$

2 Do these mentally. Write the answers only.
   a 15×10 b 1.5×10 c 1.5×100
   d 0.15×1000 e $15 \div 10$ f $1.5 \div 10$
   g $15 \div 100$ h $15 \div 1000$

3 Multiply.
   a 1.5×6 b 0.12×7 c 0.2×09
   d 1.2×0.8 e 0.05×0.2 f 13.2×12.88

4 Divide.
   a $2.46 \div 3$ b $0.255 \div 3$ c $1.755 \div 1.3$
   d $44.2 \div 0.17$ e $89.11 \div 1.9$ f $9.12 \div 3$
   g $30.69 \div 3.1$ h $61.48 \div 2.9$

5 What is the cost of 13 litres of cooking oil at $5.50 per litre?

6 A rocket travels at 88.5 m per second. How far will it travel in 25.5 seconds?

7 It takes Earth one year, or 364.25 days, to complete one revolution around the Sun. It takes Venus 0.616 Earth years to complete one revolution.
   a How many days will it take for Earth to complete five revolutions?
   b How many days does it take Venus to complete one revolution?

Topic 22 Perimeter and Area Workbook pages 69–72

▲ Imagine that you want to create a vegetable garden like the one in this picture. How much wood would you need to make the vegetable boxes and the fence? What would you have to measure? How would you do this? How would you work out how much space the vegetable boxes take?

In real life, we often need to know the dimensions of flat shapes or figures; for example, the **area** of a basketball field or the **perimeter** of a garden. In this topic, you will estimate and measure perimeters of figures; identify and describe the **circumference** and area of circles; calculate areas of regular and irregular shapes using **square units**; and learn when to calculate area and when to calculate perimeter.

Getting Started

1   Estimate the perimeter of the door in your classroom. Then measure it. How accurate was your estimate?

2   Which of the following could be measurements of area? How do you know?
    $16\,cm^2$ $100\,m^2$ $18\,cm$ $12\,m^2$

3   Can you work out the area of a circle if you know the radius? Explain what you need to do.

4   The diagram is the plan of a plot of land drawn at a scale of 1 : 200. Measure the lengths of the sides on the plan.
    a   Work out the real lengths using the scale.
    b   Calculate the perimeter of the plot of land.
    c   What is the area of the plot of land?

1 : 200

Unit 1 Measuring Perimeter

Let's Think ...

How could you calculate or measure the perimeter of these swimming pools?

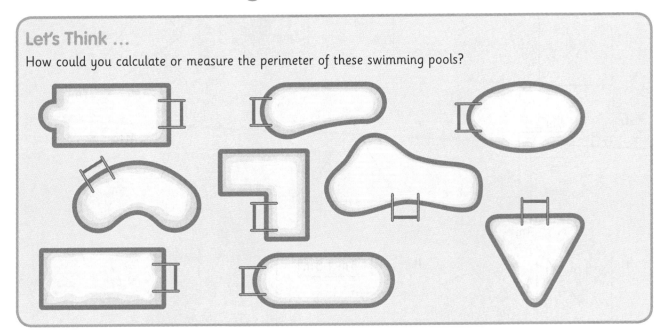

The *perimeter* of an object is the distance around it. You can estimate or guess the approximate measurement without using any measuring instruments.

To measure the actual perimeter, you need to measure the length of each side and add them up. If you have been told the lengths of the sides then just add them up.

To measure a curved side, you can lay a piece of string along the edge, then measure the string.

Use mm, cm, m or km to measure perimeter, depending on the size of the shape that you are measuring.

If the shape that you want to measure is regular, you can use formulae to calculate the perimeter.

Rectangles

The opposite sides (length and width) are equal, so:

the perimeter (P) equals twice the length (l) plus twice width (w)

$$P = 2 \times l + 2 \times w \qquad or \qquad P = 2 \times (l + w)$$

Squares

All four sides are the same length, so:

the perimeter (P) equals four times the length of one side (s)

$$P = 4 \times s$$

1   Estimate and then measure the perimeters of the following shapes in your home or at school:

    a   a window   b   a carpet

    c   a computer screen   d   a sports field

2 Find the perimeter of each shape without measuring.

a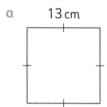
13 cm

b
11 mm
7 mm
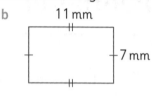

c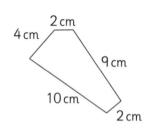
2 cm
4 cm
9 cm
10 cm
2 cm

d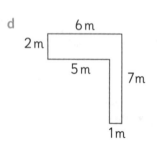
6 m
2 m
5 m
7 m
1 m

e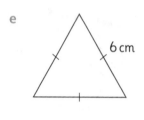
6 cm

f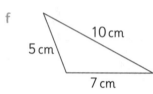
10 cm
5 cm
7 cm

g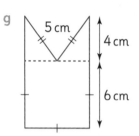
5 cm
4 cm
6 cm

h
1.5 m   1.5 m
5 m
8 m
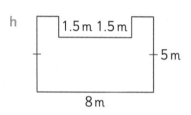

3 Draw the following shapes.

   a Two different triangles, each with a perimeter of 12 cm.

   b Two different rectangles, each with a perimeter of 24 cm.

4 Solve these problems.

   a The perimeter of a square table cloth is 5 metres. What size square table will this cloth cover?

   b The perimeter of a rectangle is 20 cm. One side is 8 cm. How long is the other side?

   c Tori-Anne wants to put patterned paper tape around the perimeter of the front of her exercise book. The book measures 15 cm × 21 cm. The tape is 1 cm wide. How many cm of tape will she need to cover the perimeter?

   d Seymour wants to put some photographs on a poster. The poster is 15 cm × 30 cm. He wants to put at least 6 photographs on the page. He wants a little space around each photograph. What should the perimeters of the photographs be?

Looking Back

1 Which of these is the formula for calculating the perimeter of a rectangle:
$P = 2 \times (l + w)$ $P = 4 \times l + 4 \times w$

2 Write the formula for calculating the perimeter of a square.

3 Calculate the perimeters of shapes with these measurements.

   a An equilateral triangle with sides measuring 7 cm.

   b A square with sides measuring 11 cm.

   c A rectangle that measures 2.5 m × 3 m.

   d A pentagon with the following measurements: 12 mm, 11 mm, 15 mm, 15 mm and 14 mm

Unit 2 Measuring Area

Let's Think ...

- What do you need to measure if you want to cover a bathroom wall with tiles? How can you work out how many tiles you need?
- Find the areas of the shapes below by counting the number of 1 cm squares. Write down your answers like this ___ cm².

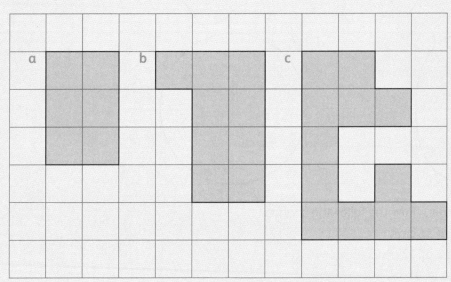

Area is a measure of how much space a flat shape takes up. You measure area in *square units* such as *cm². One square centimetre (1 cm²) is a square with a length and width of 1 cm. One square metre is a square with a length and width of 1 m. You calculate area by counting square units or by multiplying.*
You can use formulae to calculate the area of polygons.

Rectangles

The area (A) equals the length (l) times the width (w):

   $A = l \times w$

Squares

The sides of a square are equal, so the area (A) equals, the length (l) times the length (l):

   $A = l \times l$ *or* $A = l^2$

Right-angled triangles

You can also use a formula to work out the area of a right-angled triangle.
The area (A) equals half the length of the base (b) times the height (h):

   $A = \frac{1}{2} \times (b \times h)$

1 Estimate the area of these irregular shapes. Discuss what to do with the squares that are partly covered. Compare your estimates.

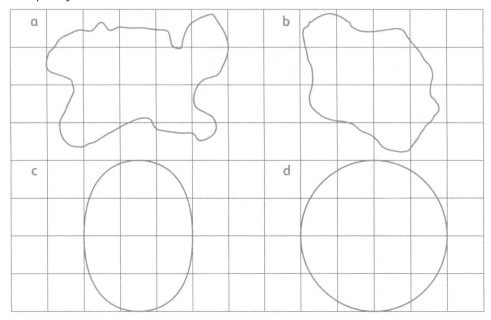

2 Use centimetre square paper. Draw the following shapes.

  a A square with an area of 16 cm².

  b A rectangle with an area of 12 cm².

  c Any shape with an area of 9 cm².

3 Use a formula to work out the areas of these rectangles and squares.

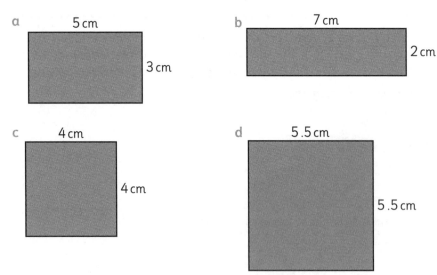

4 Calculate the area of the following shapes.

  a A wall that measures 2.2 m by 3.6 m.

  b A book cover with a perimeter of 70 cm and a width of 15 cm.

  c A billboard with a perimeter of 24 m and a height of 4 m.

  d A mobile phone with a width of 6.5 cm and a height that is twice the length of the width.

5 Estimate the area of the triangles below by counting the squares. Then use a formula to check your estimates.

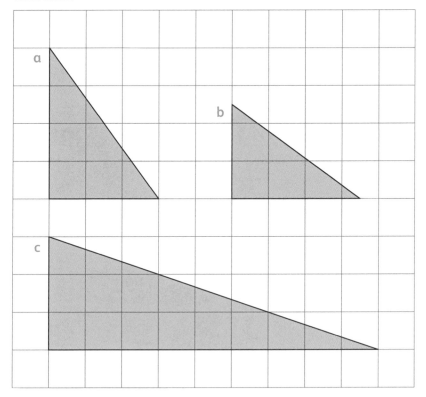

6 How could you work out the area of this floor?

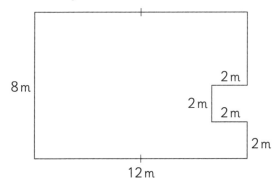

Looking Back

1 Write down the following formulae.
   a A formula to calculate the area of a square.
   b A formula to calculate the area of a rectangle.
   c A formula to calculate the area of a right-angled triangle.
2 Work in pairs. Make up problems involving area for your partner to solve.

Unit 3 Measuring Circles

Let's Think ...

- Use a pair of compasses to draw and then cut out circles with:
  a radius of 4 cm   a radius of 5.5 cm   a radius of 6 cm.
- How can you work out the diameter of each circle? How can you check your calculation?

The perimeter of a circle is called the circumference. The distance from the centre point of the circle to any point on the circumference is called the radius.

A diameter is any straight line that passes from one point on the circumference, through the centre point, to the circumference on the other side. The length of the diameter is always twice the length of the radius, so d = 2 × r.

You can use pi (π) to calculate the circumference (C) of a circle:

$$C = π × d \quad \text{or} \quad C = 2 × π × r$$

You can calculate the area (A) of a circle using this formula:

$$A = π × r^2$$

Pi (π) = 3.14 or $\frac{22}{7}$

Calculate the circumference (perimeter) and area of this circle.

Circumference = 2 × π × r	Area = π × r²
= 2 × 3.14 × 4	= 3.14 × 4 × 4
= 25.12 cm	= 50.24 cm²

Remember, area is measured in square units.

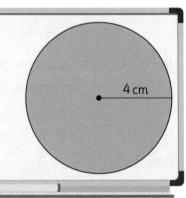

4 cm

1. Work out the circumference and the area of the following circles.
   a A circle with a radius of 3 cm.
   b A circle with a diameter of 30 mm.
   c A circle with a radius of 2.6 cm.
   d A circle with a diameter of 1.2 m.
   e The combined circumferences of two circles, each with a radius of 25 cm.

2. Discuss how you could work out the diameter of a circle if you know what the circumference is.

3. You draw a circle with a radius of 10 cm. Is the circumference greater than or smaller than 80 cm?

Looking Back

1. Name one measurement that you need in order to be able to work out the circumference of a circle.
2. Is the circumference of a circle its area or its perimeter?
3. What can you calculate with the following formulae?
   a $A = π × r^2$
   b $C = 2 × π × r$

Topic Review

What Did You Learn?

- The perimeter is the distance around the edge of a figure or shape.
- Use mm, cm, m or km to measure perimeter, depending on the size of the shape that you want to measure.
- You can calculate perimeter by adding up the lengths of each side of a figure.
- The formula for the perimeter of a rectangle is twice length plus twice width, or $P \times 2 \times l \times 2w$ or $P \times 2 \times (l \times w)$.
- The formula for the perimeter of a square is four times the length of one side, or $P = 4 \times s$
- The circumference of a circle is the measure of the distance around the circle or its perimeter.
- The circumference C is equal to π times the diameter d or $C = \pi \times d$ or $C = 2 \times \pi \times r$
- Pi $(\pi) = 3.14$ or $\dfrac{22}{7}$
- Area is the amount of space a flat figure takes up, measured in square units.
- To calculate the area of a rectangle, use the formula: $A = l \times w$
- To calculate the area of a square, use the formula: $A = l \times l$ or $A = 2 \times l^2$
- To calculate the area of a right-angled triangle, use the formula: $A = \dfrac{1}{2} \times (b \times h)$

Talking Mathematics

Explain what these words mean.
- Circumference
- Radius
- Diameter
- Area
- Perimeter

Quick Check

1 Work out the perimeter and area of each of these shapes.

a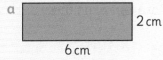
2 cm
6 cm

b 3 cm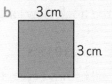
3 cm

c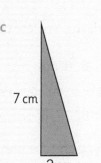
7 cm
2 cm

d
9 mm

2 Explain in your own words the main difference between calculating the perimeter of a square and the perimeter of a rectangle.

Topic 23 Speed, Distance and Time

Workbook
pages 73–74

▲ What unit of measurement do we typically use for the distance that a car travels? Do you know how we typically measure the speed of a car? How would you know at what speed a car is driving at any given moment?

When we talk about our heart rate, we use the term 'beats per minute'. The word 'per' is a clue that we are talking about a rate. A **rate** compares two quantities in different units, such as dollars per hour as a rate of pay. In this topic, you will learn how distance, time and rate are related when we want to calculate the **speed** of an object. You will also learn how to apply formulae to calculate **distance, time** and speed.

Getting Started

1   A grocer sells half a dozen eggs for $1.80. A supermarket down the road sells a dozen eggs for $3.90. Which store offers the best price rate per egg?

2   What is your heart rate in beats per minute? Find a way of measuring it and compare your rate with a partner.

3   If Mr Sands travels 168 km in two hours, what is his rate of travel in kilometres per hour? How did you work this out?

4   A runner takes 20 seconds to complete a lap of the track. How long will it take her to run these distances at the same speed?

   a   2 laps   b   6 laps   c   10 laps

5   Terrold cycles 12 kilometres in an hour. How far will he have cycled in:

   a   2 hours   b $3\frac{1}{2}$ hours   c   0.5 hours

6   Ms Simpson is driving along and she sees a sign with the number 60 on it. What does the sign mean?

Unit 1 Speed as a Rate of Distance and Time

Let's Think …

- If it takes you half an hour to walk three kilometres, how far will you go if you continue walking for an hour?
- A car covers a distance of 180 kilometres in two hours. What distance did it cover in one hour if it travelled at the same speed throughout?
- Stefan rides his bike at a constant speed of 10 kilometres per hour. How long will it take him to travel a distance of 15 kilometres?

Speed is the rate of distance per unit of time. You can say that a car is driving at 60 kilometres (distance) per hour (time), or 60 km/h. This means, if the car continues at the same speed for one hour, it will cover 60 kilometres. When athletes run the 100 m sprint, their speed is measured in metres per second.

Sometimes, the speed of an object is given as a rate that you can work out quickly. An airplane that covers a distance of 860 km in one hour has an average speed of 860 km/h. A car that drives 50 km in half an hour travels at a speed of 100 km/h.

Some rates are more complicated to work out; for example, what is the speed in km/h of a cyclist who covers 23 km in 45 minutes?

To calculate the speed of a travelling object, you use the formula or rule:

$$Speed = \frac{Distance}{Time}$$

Example 1

A truck leaves the harbour at 4 : 10 p.m. and arrives at the warehouse at 5 : 40 p.m. The warehouse is 84 km from the harbour. Did the truck driver exceed an average speed of 60 km/h?

You can use the formula to calculate the truck's average speed. Remember that the time must be in hours and not in minutes.

$$Speed = \frac{84 \text{ km}}{1.5 \text{ h}} = 56 \text{ km/h}$$

The truck travelled at an average speed of 56 km/h and therefore did not exceed 60 km/h.

You can also use this relationship between distance, time and speed to determine the distance travelled when you know the speed and the time:

$$Distance = Speed \times Time$$

Example 2

Darrius walked from his home to his grandparents' home at a steady speed of 5 km/h. He walked for 3 hours. How far did Darrius walk?

Use the formula:

$$Distance = Speed \times Time$$
$$= 5 \text{ km/h} \times 3 \text{ h} = 15 \text{ km}$$

Darrius walked 15 km in 3 hours.

To work out the time taken when you know the distance and speed, the formula looks like this:

$$Time = \frac{Distance}{Time}$$

Example 3

Trudy cycles at 8 km/h and covers a distance of 20 km. How long does her journey take?

$$Time = \frac{20 \text{ km}}{8 \text{ km/h}}$$

$$= 2\frac{4}{8}\text{h}$$

$$= 2\frac{1}{2}\text{h}$$

$$= 2.5\text{h}$$

Trudy's journey took 2 hours 30 minutes.

This triangle diagram shows a quick way of remembering the formula for calculating speed, distance and time.

$$D = S \times T$$
$$S = \frac{D}{T}$$
$$T = \frac{D}{S}$$

1 Tyrone drives for 400 km at an average speed of 80 km/h. How long did he drive for?

2 Bernadette runs from 10:40 a.m. until 11:10 a.m. at an average speed of 9 km/h. How far did she run?

3 Perry ran 4 500 m in half an hour and Jerome ran 6 000 m in 40 minutes. Who ran the fastest?

4 What is the speed of a train that travels 243 km in 2 h 15 min?

5 Melanie spent 4 hours on a plane. For half of that time, the plane's average speed was 900 km/h and for the rest of the time, its average speed was 760 km/h. What distance did Melanie travel?

Looking Back

1 What would you mean if you say that you are travelling at 80 km/h?

2 How do you calculate speed, distance and time?

3 What units of measurement do you use for speed?

4 What do you need to do if the units of measurement are not the same?

Topic Review

What Did You Learn?

- Speed is the rate of distance per time. Speed can be expressed as kilometres per hour, or km/h.
- The formula for calculating the average speed of an object when you know the distance it has covered and the time spent travelling is given by

$$\text{Speed} = \frac{\text{Distance}}{\text{Time}}$$

- The formula for calculating the distance an object has travelled when you know its average speed and the time it took is given by

$$\text{Distance} = \text{Speed} \times \text{Time}.$$

- The formula for calculating the time it has taken an object to cover a certain distance at a certain speed is given by

$$\text{Time} = \frac{\text{Distance}}{\text{Speed}}$$

Talking Mathematics

Give the mathematical term for each of the following.

- A ratio that compares two different kinds of numbers and usually contains the word 'per'.
- Distance per time.
- Distance per speed.
- Speed times time.

Quick Check

1 What distance would you travel if you drove for 4 hours at 80 km/h?

2 How long does it take to cycle 35 km at an average speed of 7 km/h?

3 What is the average speed of a train that covers 340 km in 2 hours?

4 Mr Davis travelled 450 kilometres in 5 hours.

   a What was his speed in km/hr?

   b How far would you expect him to travel in $7\frac{1}{2}$ hours if he kept to the same speed?

   c How long would it take to cover 1 080 km at this speed?

Topic 24 Probability Workbook page 75

▲ Most dice are cubes. Have you seen any other kinds of dice? How many sides do they have? Compare the dice in the photograph. Do you still have an equal chance of rolling each number on the dice?

When you roll an ordinary die, there are several possible **outcomes**. You have an equal chance of rolling any of the six numbers. **Probability** is the calculation of **likely** outcomes. In this topic, you will use probability to predict how likely things are to happen. You will also work out what the **possible** outcomes are in different situations and then work out how likely each outcome is. You will also learn how to describe a **fair** game.

Getting Started

1 What do the words impossible, unlikely, likely and certain mean when you are talking about probability?

2 Predict how likely these things are. Sort them into four categories: impossible, unlikely, likely, certain.

- The sun will rise tomorrow.
- There will be snow outside our classroom tomorrow.
- The next Olympics will take place in The Bahamas.
- It will rain sometime tomorrow.
- There will be six moons in the sky tonight.
- You will go to bed after midnight tonight.
- You will do some homework this afternoon.
- One hour after 7 o'clock it will be 8 o'clock.

3 Add one more event of your own to each category.

Unit 1 Calculating Probabilities

Let's Think …

- When you spin the arrow for Spinner A, which colours could it land on?
- What fraction of Spinner A does each colour take up?
- When you spin the arrow for Spinner B, which colours could it land on?
- What fraction of Spinner B does each colour take up?
- Which spinner gives you the best chance of landing on red?

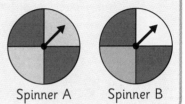

Spinner A   Spinner B

Probability tells you how likely something is to happen. If something is certain to happen, the probability is 1 or 100%. When there are several possible outcomes, you use probability to predict the likelihood of each outcome.

When you throw a normal 6-sided die, there are six possible outcomes: 1, 2, 3, 4, 5 and 6. You have an equal chance of each outcome.

The probability of throwing a 3 is 1 out of 6, or $\frac{1}{6}$

$P(3) = \frac{1}{6}$.

The probability of throwing any number that is not 3 (in other words 1, 2, 4, 5 or 6) is 5 out of 6 or $\frac{5}{6}$.

$P(not\ 3) = \frac{5}{6}$

Probability of a favourable outcome = $\dfrac{Number\ of\ favourable\ outcomes}{Total\ number\ of\ possible\ outcomes}$

An impossible outcome has a probability of zero or 0%; for example, when you throw an ordinary 6-sided die, there is a zero possibility of throwing a 7.

1   Look at Spinner A above. Express as a fraction the probability of landing on:

    a red   b green   c any colour other than red or green

2   Look at Spinner B above.

    a How many possible outcomes are there?

    b List the possible outcomes, and express the probability of each as a fraction.

3   When you roll a fair 6-sided die, what is the probability of throwing:

    a 6   b an even number

    c an odd number   d a number under 7?

4   Shaundra has a bag with 20 counters in four different colours: red, blue, yellow and purple. The probability of choosing each colour if she pulls out a counter without looking is given in the table. Use this information to work out how many of each colour counter there are in the bag.

Colour	Probability
Red	$\frac{1}{4}$
Blue	$\frac{1}{5}$
Yellow	$\frac{1}{10}$
Purple	$\frac{9}{20}$

Tree Diagrams

A tree diagram helps you to organize the outcomes of an experiment.

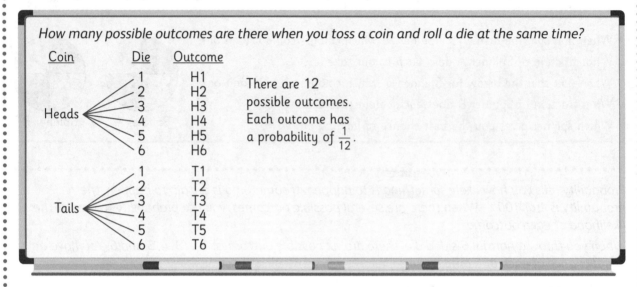

How many possible outcomes are there when you toss a coin and roll a die at the same time?

Coin	Die	Outcome
Heads	1	H1
	2	H2
	3	H3
	4	H4
	5	H5
	6	H6
Tails	1	T1
	2	T2
	3	T3
	4	T4
	5	T5
	6	T6

There are 12 possible outcomes. Each outcome has a probability of $\frac{1}{12}$.

5 Nathan writes the letters of his name on pieces of paper and puts them all in a bag. He draws out a letter at random. The tree diagram shows all the possible outcomes.

  a What is the probability of drawing an H?

  b What is the probability of drawing an N?

  c Which letters have an equal probability of being drawn?

N
A
T
H
A
N

6 Glenroy has grey shorts and blue shorts and red, blue, black and white T-shirts.

  a Draw a tree diagram to show all the possible combinations he could choose to wear.

  b What is the probability that the shorts and T-shirt will be the same colour?

7 Toniqua drew this tree diagram to show her lunch options at a beach cafe. List all the possible combinations she could have for lunch.

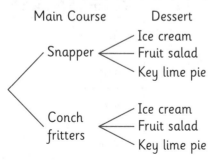

Main Course   Dessert

Snapper
  Ice cream
  Fruit salad
  Key lime pie

Conch fritters
  Ice cream
  Fruit salad
  Key lime pie

8   Work out whether each of these games are fair.

  a   If the die lands on 1, 2, or 3, Matthew gets a point. If the die lands on 4, 5 or 6, Alex gets a point.

  b   If the die lands on a 6 or a 1, Mathew gets a point. Otherwise, Alex gets a point.

9   If the arrow lands on black, Maya gets a point.
    If it lands on red, Theresa gets a point.
    If it lands on yellow, both players get a point.
    Is the game fair?

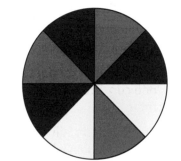

10   Rashad offers Silly Billy this game: 'You roll the die. If you get a six, I pay you $2.00. If you do not get a 6, you pay me $2.00.' Should Silly Billy accept? Why or why not?

11   Make up your own Silly Billy games using a die. Make at least two unfair challenges and one fair challenge.

12   Try this challenge. Mike and William play a game where they throw two dice and add up the numbers to get a score. Mike gets a point if the sum is 7 or lower, and William gets a point if the sum is 8 or higher.

  a   What are the possible outcomes of throwing two dice? (Hint: work out the possible combinations, and the sum of each.)

  b   Work out whether the game is fair or unfair.

13   Sybil and Marianne roll a die to decide who goes first in a game. Which of these methods are fair? If a method is not fair, say why not.

	Sybil	Marianne
Method 1	Must roll an odd number	Must roll an even number
Method 2	Must roll 3	Must not roll 3
Method 3	Must roll less than 3	Must roll more than 3
Method 4	Must roll a prime number	Must roll a composite number

Mixed Probability

Work through these problems in pairs.

1 For each sentence, describe the probability using words such as likely, unlikely, certain, impossible, possible.

   a Today is Wednesday, and tomorrow will be Thursday.

   b Today is Sunday, and tomorrow will be Wednesday.

   c If I ran a sprint against Usain Bolt, he would win.

   d The sun will set at 11:00 a.m. tomorrow.

   e If I flip a coin, it will land on heads.

   f Tomorrow it will rain.

2 a If an outcome is certain, it has a probability of ☐.

   b If an outcome is impossible, it has a probability of ☐.

   c If there is a 9 out of 10 probability of rain today, what is the probability that it will not rain today?

3 Write the possible outcomes of:

   a flipping a coin

   b throwing a normal six-sided die.

4 An ice-cream vendor sells vanilla, strawberry, chocolate, coconut and lime flavours. You can choose a cone or a cup. Draw a tree diagram to show all the possible combinations you could choose. How many are there?

5 Rashad offers Silly Billy this game: 'If the arrow lands on yellow, you get $5.00. If it lands on red, I get $5.00.' Explain whether Silly Billy should accept the challenge or not.

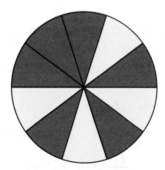

Looking Back

1 TRUE or FALSE? If false, explain why.

   a When you flip a coin, you are more likely to get heads than tails.

   b When you throw a die, you have even chances of landing on an odd or an even number.

   c When you throw a die, you have a 1 in 3 chance of getting a multiple of 3.

2 Micah, Glenroy, Sandra and Toniqua are at the waterpark. Only two of them can go on the slide at the same time. How many combinations of two students are possible? List them all.

Topic Review

Talking Mathematics

How can a tree diagram help you work out the probability of a particular outcome?

Quick Check

1 Copy and complete each sentence correctly to describe this spinner.

   a The arrow has a (1 in 3/ 1 in 6/ 50 %) chance of landing on red.

   b The arrow is (more likely/less likely/equally likely) to land on grey as to land on red.

   c It is (possible/impossible/likely/unlikely) that the arrow will land on black.

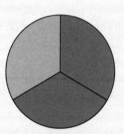

2 Answer the questions about this spinner.

   a On which colour is the arrow most likely to land?

   b On which colour is the arrow least likely to land?

   c Is it impossible for the arrow to land on yellow?

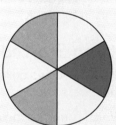

3 Zahria and Kayla have a deck of cards. They play this game:

   If a number card from 2 to 7 is dealt, Zahria gets a point. If a number card from 8 to 10, or a picture card (jack, queen, king) is dealt, Kayla gets a point. If an ace is dealt, they both get a point. The first person to get 10 points wins.

   a Is it a fair game?

   b Why is it important to shuffle the deck properly first?

Topic 25 More Measures Workbook pages 76–77

▲ What is this woman measuring? What unit of measurement is she using?
How accurate is this measurement? Is this measurement still used today?

We can measure everything, from time, temperatures, mass and distances to light and energy. Over time, people have developed different ways of measuring and different **units of measurement**. As there is a lot more communication and trade between people all over the world nowadays, standard units and formats of measurement have been developed to avoid confusion. You are going to learn more about these units and formats in this topic.

Getting Started

1 If you have two gallons of water and your friend has two litres of water, who has more water? What is the difference?

2 Which countries measure mass and money in pounds? How would they write a mass of 20 pounds and a price of 20 pounds?

3 What would be a more accurate way of writing 6 o'clock? Why?

4 What information do you have to give when you fill in a part of a form that looks like this:

Y	Y	Y	Y	M	M	D	D

Unit 1 Different Ways of Measuring

Let's Think ...

- Which of these units could you use to measure mass?
- Which units are metric and which are customary units?
- Which units would you use to measure a large mass?

pound	gallon	kilogram	gram
stone	ounce	milligram	

In different countries of the world, people measure things using different units of measurement. Sometimes, people use customary units such as stones, pounds and ounces to measure mass.

Many countries use kilograms to measure mass. This is a metric measure which was first used in France in 1799. At that time, it was decided that one kilogram should be equal to the mass of one litre of water at 4°C. Later, scientists decided to make a kilogram in metal to use as a standard. This kilogram is kept under three bell jars in a specially guarded place in France. There are copies of this kilogram in other countries of the world.

▲ This is a copy of the kilo prototype which is kept in the USA.

Sometimes, people use customary units such as feet, inches, yards and miles to measure length or distance. The USA, Myanmar and Liberia are the only countries who still officially use these measurements. All other countries have officially changed to the metric system, even though some people still talk about and use customary measures.

Money is a form of measurement. The amount of money you have can be written in different ways. Some people use spaces, others use commas and others use decimal points to write amounts of money. Here are some examples:

$15,00 $15.00 $1 500.00 $1,500.00

1 Work in pairs. Find the answers to the following questions. Then compare your answers.
  a What is the customary system of measurement?
  b Why is the metric system used in so many countries?
  c If a person weighs 8 stones, how much do they weigh in kilograms?

2 Work in pairs. Which person is the tallest? You will need to find out how to convert feet and inches to metres in order to work this out.

I am five foot four

I am one point four metres

I am six foot three

3 Work in groups. Make a list of five countries that use dollars as a currency. Then find out how they would write the following amounts in numbers. Draw a diagram or make a table to show this comparison.
  a Twenty dollars fifty cents
  b One thousand thirty dollars

Work in groups to discuss and solve these problems.

4 Imagine that you are farmers or traders. There is no common unit of measuring mass. You want to trade the goods that you grow or produce. What could you use as a unit of measurement? What would you need to think about?

5 You have a factory that produces potato chips. You want to change the packaging of the chips. To do this you need to work out how many chips (more or less) could fit in each bag. How would you do this?

6 Your school is preparing for a carnival. Students from several other schools will be coming to your school to join in the celebrations. You need to work out how many students can stand on one of the school sports field at the same time. The students need to be able to swing their arms without hitting each other. How could you work this out?

7 Look at this graph carefully.
  a Explain what this graph shows.
  b Use the graph to convert 40 kilometres to miles.
  c How many kilometres is 25 miles?
  d Which is further, 60 km or 45 miles?

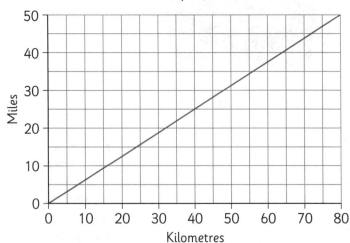

Conversion Graph: kilometres–miles

Looking Back
Write a paragraph to explain the difference between metric units and customary units.

Unit 2 Standard Formats for Time and Dates

Let's Think …

Look at these dates. We can write the date 6 November 1916 as:

> 6/11/16 or 11/6/1916 or 1916.11.06.

- Why could this be confusing?
- Which way would you write this date?

The International System of Units (SI) is based on the metric system. It uses standard measures such as metres, kilograms and seconds to measure.

Standard formats are often used for recording dates and times in order to avoid confusion. An organization, called the International Standards Organization (ISO), sets the standard formats for recording times and dates.

There are several advantages to this:

- *Data capturing is easier.*
- *The dates are more easily recognized by computer software.*
- *You do not need to be able to understand another language in order to understand these dates.*

Dates in Standard Format

To write the date 12 May 2017 in standard format, first write the year, then write the month and then write the day:

> *2017-05-12*   *(YYYY-MM-DD)*

You should put a zero in front of days or months with single digits (1–9).

Year				Month		Day	
2	0	1	7	0	5	1	2

Time in Standard Format

To write the time 15 minutes and 55 seconds past nine in the evening using standard format, first write the hour, then the minutes and then the seconds:

> *21:15:55*   *(HH:MM:SS)*

You should always use the 24-hour clock when using standard format.

---

1   Write your birthday in standard format.

2   Match the following dates.

1866-11-02	11th February 1866
2010-10-03	12th August 1942
1942-08-12	3rd October 2010
1866-02-11	10th March 2010
1942-12-08	2nd November 1866
2010-03-10	8th December 1942

3 Write each date in standard format.

  a 15 June 2016

  b 18 December 2017

  c January 9th, 2012

  d February 16th, 2020

4 Work in groups. Discuss and then do further research to find out when standard formats are used for dates; for example, how would you write your date of birth if:

- you filled in a form to apply for an identify document?
- you filled in a form to open a bank account?
- you filled in a form because you wanted to join a sports club?

5 Work in pairs. Use your social science textbook to find 10 dates that you think are very important in the history of your country. Make a timeline of the dates, writing each date in ISO format.

Looking Back

1 Give two reasons why it is useful for people to write dates and times in a standard format.

2 Look at these dates of birth. On which day of which month were these people born?

  a 2001-04-19

  b 2002-12-07

  c 1999-09-09

  d 2002-02-02

Topic Review

Talking Mathematics

Use the internet to find out what these units of measurement are and what they measure. Explain what the metric equivalent of each is.

- furlong
- fortnight
- firkin
- a light year
- a dog year

Quick Check

1 Write the following birthdays in ISO format.

a 12th July 2006

b 4th January 2000

c 01 May 1954

d 27th June 1999

2 Read the following times aloud in as many ways as you can.

a 15:30

b 21:55

c 09:30:16

d 12:01:00

Topic 26 Looking Back

Revision A

1 Write the following in numerals.
  a Three billion sixty-four million twenty-five thousand two hundred three
  b Ninety-nine thousand, four hundred sixty-three

2
| 4, 17, 24, 10, 41, 49, 12 |

  From the set of numbers in the box, write down:
  a a multiple of 8
  b two square numbers
  c a prime number
  d a factor of 8
  e the least common multiple of 6 and 4
  f a triangular number.

3 Write these numbers in words.
  a 12 312 065
  b 234 100 876
  c 9 452 345 098

4 Write down an example of where you might come across each of these types of numbers in your daily life.
  a Positive integer.
  b Fraction.
  c Percentage.
  d Negative integer.
  e Decimal.

5 If each block on this diagram represents a square with sides of 1 cm. What is the area covered by the drawing?

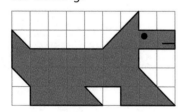

6 Write down the next three numbers in each of these number patterns.
  a 3, 6, 9, 12, 15, …
  b 10 000, 1 000, 100, …
  c 0, 1, 3, 6, 10, …
  d −12, −10, −8, −6, …

7 Choose the correct name from the box for each angle.

| Acute | Obtuse | Right |
| Straight | Reflex | Revolution |

  a   b   c

8 Calculate.
  a $3 \times 7 + 3 \times 6$ b $7\,654 \times 0$
  c $36\,456 \times 100$ d $876 - 498$

9 Use rounding to estimate the answers to these calculations.
  a $603 + 715 + 986$
  b $7\,899 - 5\,211$
  c $24\,999 \div 2$
  d 408×31

10 Describe the probability of each of these events.
  a Monday will follow Sunday.
  b Snow will fall in Nassau tomorrow.
  c Rolling a six on a normal die.
  d You will beat an Olympic gold medallist over 100 m.
  e It will be cold in Antarctica.
  f A new light bulb will last longer than two weeks.

11 Simplify each fraction.
  a $\dfrac{33}{3}$ b $\dfrac{11}{2}$
  c $\dfrac{25}{60}$ d $\dfrac{8}{100}$

12 Change each mixed number to an improper fraction.

   a $7\frac{1}{5}$
            b $8\frac{1}{3}$

13 Calculate and give the answers in simplest form.

   a $1\frac{1}{2}+2\frac{2}{3}$
        b $6-4\frac{2}{3}$

   c $\frac{4}{5}\times\frac{2}{6}$
        d $4\times2\frac{3}{4}$

Revision B

1 Change each decimal to a fraction or mixed number in simplest form.

   a   0.3
          b   0.45

   c   4.25
          d   5.44

2 Calculate.

   a   1.9 + 2.43
      b   18 − 2.75

   c   0.2 × 0.02
     d   1.84 × 10

   e   23.45 × 100
     f   7.24 ÷ 4

   g   245 ÷ 100
      h   3.8 ÷ 0.2

3 Round these numbers.

   a   0.786 to the nearest tenth

   b   194.123 to the nearest hundredth

   c   19.15567 to two decimal places

   d   0.999 to two decimal places

4 Look at the shape below.

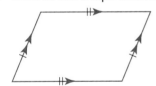

   a   What is the mathematical name for this shape?

   b   How many vertices does this shape have?

   c   How many angles does it have?

   d   How many pairs of parallel sides does it have?

5 Write each percentage as a fraction or mixed number in simplest form.

   a   70 %
        b   17 %

   c   140 %
      d   2.5 %

6 Write as a percentage.

   a $\frac{1}{2}$
        b $\frac{4}{5}$

   c $1\frac{1}{4}$
        d $\frac{48}{200}$

7 Write each percentage as a decimal fraction.

   a   85 %
        b   5 %

   c   25.5 %
      d   200 %

8 Calculate mentally.

   a   40 % of 200
      b   6 % of 1 200

   c   23 % of 48 000
     d   99 % of 500

9 Calculate the area of each of these shapes.

10 Look at this drawing of an ant. It has been drawn at a scale of 5 : 1.

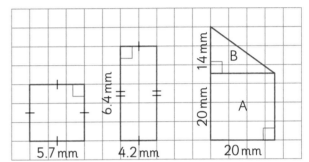

   a   What does a scale 5 : 1 mean?

   b   Is the ant in the picture bigger or smaller than a real ant?

   c   Measure the length of the ant on the drawing.

   d   Use the scale to work out how long the real ant would be.

11 Which two numbers do not belong in this set of integers? Why?

    −3   14 $\frac{3}{5}$ −9   0   −3.4

12 For each of the following shopping lists:

    a estimate the total cost

    b estimate how much change you would get from $100.00

    c work out the exact answers.

List A	List B
1 × $4.10	$56.09
4 × $0.97	2 × $2.20
2 × $3.15	$2.76
1 × $6.59	$1.95
	10 × $1.06

List C	List D
3 × $3.79	6 × $2.47
2 × $1.49	$2.75
$1.98	$4.70
4 × $1.54	4 × $0.99
	$3.80

Revision C

1 Write each of these lengths in centimetres.

    a 68 mm   **b** 108 mm

    c 3 metres   **d** a kilometre

2 Write these times as 24-hour times.

    a 43 minutes past 7 in the morning

    b 11 : 52 p.m.

    c 4 o'clock in the morning

    d 4 o'clock in the afternoon

3 How many grams are there in each of these amounts?

    a 5 kilograms

    b 130 milligrams

    c 5.5 kilograms

    d 8 700 milligrams

4 Convert each measurement to the units given.

    a 4.234 litres to millilitres

    b 7.34 litres to millilitres

    c 5 465 millilitres to litres

    d 4 300 millilitres to litres

    e 45 millilitres to litres

5 Sharyn buys 47 metres of fabric at $2.50 per metre. She sells it to her friends for $4.99 per metre. How much money does she make?

6 Study these triangles carefully.

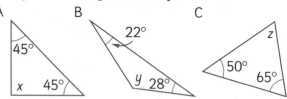

    a Choose two names for each triangle from the box below:

Scalene	Isosceles	Equilateral
Right-angled	Acute	Obtuse

    b Write equations and solve them to find the size of angles x, y and z.

7 Which units would you use to measure each of these areas?

    a The top of a desk.

    b The floor of a shopping centre.

    c This page.

    d A postage stamp.

    e The USA.

8 Students were asked how many brothers and sisters they had. Their answers are given here:

1	4	0	1	3
6	2	1	1	1
2	3	2	1	1
2	3	5	1	1
2	2	2	1	2

Draw a frequency table to organize the results.

9 Find the mean, median, mode and range of the following sets of data. Give your answers correct to one decimal place.

    a 2, 4, 2, 7, 3, 5, 4, 2, 3, 1

    b 40, 20, 30, 60, 50, 10

10 The graph below shows information about the temperature and rainfall at a weather station in The Bahamas. The bars show the rainfall and the line shows the temperature.

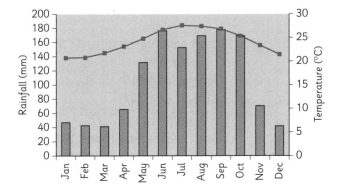

a   Why does this graph have two different vertical scales?

b   Which month has the highest rainfall?

c   Which month is the coldest?

e   What happens to the temperature from March to November? How can you tell this by looking at the graph?

11 A teacher wanted to know if the students in Grade 6 had read the newspaper over the weekend. He asked some students and drew this graph of his results.

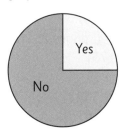

a   What kind of graph is this?

b   What fraction of the students had not read the newspaper?

c   What fraction of the students had read the newspaper?

d   If the teacher asked 40 students, how many of them had read the newspaper?

12 What do these words mean in mathematics?

a   Never   b   For sure

c   No chance   d   Maybe

13 Calculate:

a $432\,876 \div 23$

b $32\,456 \times 129$

c $124\,567 + 45\,987$

d $342\,098 - 12\,987$

14 A lion can run at $\frac{3}{5}$ of the speed of a cheetah. If a cheetah's speed is 115 km/hr, how fast can the lion run?

15 A bar of gold with a mass of 16.6888 kg is sliced into 8 equal pieces. What is the mass of each smaller piece?

16 An engineer collected information about accidents at the factory where she works. This is her data.

Year	2006	2007	2008	2009	2010
No. of Accidents	200	170	140	130	100

a   Draw a bar graph to show this information.

b   What does the graph show you about safety at this factory?

17 Write as decimals.

a $\frac{12}{100}$ b $\frac{4}{5}$

18 Mrs Jones buys 300 m of wood at $40.78 per metre. What is the total cost?

19 Linda paid $156 for 400 plastic containers. How much was each container?

20 A person is paid $20.67 per hour. If she works 38.3 hours, how much will she earn?

Key Word Reference List

The key words that you learned this year are listed here in alphabetical order. If you cannot remember the meaning of a word, turn to the page number that is given next to the word. Read the definition and look at the pictures or examples to help you remember what the word means.

acute angle (page 41)
acute triangle (page 45)
add (pages 79, 117)
approximate (page 53)
area (page 137)
bar graph (page 71)
billions (page 7)
brackets (page 117)
calculate (page 1)
capacity (page 19)
Celsius (C) (page 19)
centimetre (cm) (page 19)
centre (page 47)
check (page 84)
circle (page 96)
circle graph (page 71)
circumference (pages 47, 96, 140)
clustering (page 53)
common factor (page 61)
common multiple (page 61)
compare (pages 27, 94)
compatible (page 79)
composite number (page 13)
congruent (page 109)
create (pages 90, 133)
customary units (pages 21, 153)
data (page 74)
day (page 53)
decimal (pages 35, 125)
decimal places (page 131)
decimal point (pages 35, 127)
decrease (pages 25, 31)
denominator (page 93)
describe (page 111)
diameter (pages 47, 96, 140)
difference (page 79)
digit (pages 5, 51, 129)
distance (page 143)
divide (pages 55, 101, 117)

dividend (pages 104, 132)
divisor (page 104)
double (page 57)
double bar graph (page 69)
edge (page 140)
equation (page 88)
equilateral triangle (page 45)
equivalent (pages 25, 37, 121)
estimate (pages 1, 53, 79, 134)
expression (page 88)
extend (page 10)
factor (page 61)
Fahrenheit (F) (page 19)
fair (page 149)
flip (page 112)
foot (page 153)
fraction (pages 25, 67, 121)
frequency table (page 74)
gallon (page 21)
geometric (page 12)
gram (g) (page 19)
greatest common factor (GCF) (page 61)
grouping symbols (page 117)
half (page 57)
hour (page 19)
hundredth (page 35)
image (page 111)
inch (page 153)
increase (page 31)
integer (page 31)
inverse (pages 88, 101)
irregular shapes (page 133)
isosceles triangle (page 45)
jotting (page 57)
kilogram (kg) (page 152)
kilolitre (kL) (page 19)
kilometre (km) (page 19)

least common multiple (LCM) (page 61)
length (page 19)
likely (page 147)
line graph (page 71)
line of symmetry (page 109)
line symmetry (page 109)
litre (L) (page 19)
map (page 67)
mass (pages 19, 152, 155)
mean (page 75)
median (page 75)
mental (pages 57, 99)
metre (m) (page 19)
metric measures (page 153)
mile (page 153)
milligram (mg) (page 19)
millilitre (mL) (page 19)
millimetre (mm) (page 19)
millions (page 6)
minute (page 19)
mixed number (pages 25, 121)
mode (page 75)
month (page 19)
motion (page 111)
multiple (page 61)
multiply (pages 101, 117)
multi-step (page 1)
nearest (page 51)
negative (page 31)
non-routine (page 85)
number (page 5)
numerator (page 25)
oblong number (page 13)
object (page 111)
obtuse angle (page 41)
obtuse triangle (page 45)
operations (pages 88, 117)
order (page 27)